Positron Scattering
in Gases

NATO ASI Series

Advanced Science Institutes Series

A series presenting the results of activities sponsored by the NATO Science Committee, which aims at the dissemination of advanced scientific and technological knowledge, with a view to strengthening links between scientific communities.

The series is published by an international board of publishers in conjunction with the NATO Scientific Affairs Division

A	Life Sciences	Plenum Publishing Corporation
B	Physics	New York and London
C	Mathematical and Physical Sciences	D. Reidel Publishing Company Fordrecht Boston, and Lancaster
D	Behavioral and Social Sciences	Martinus Nijhoff Publishers
E	Engineering and Materials Sciences	The Hague, Boston, and Lancaster
F	Computer and Systems Sciences	Springer-Verlag
G	Ecological Sciences	Berlin, Heidelberg, New York, and Tokyo

Recent Volumes in this Series

Volume 102—Magnetic Monopoles
edited by Richard A. Carrigan, Jr., and W. Peter Trower

Volume 103—Fundamental Processes in Energetic Atomic Collisions
edited by H. O. Lutz, J. S. Briggs, and H. Kleinpoppen

Volume 104—Short-Distance Phenomena in Nuclear Physics
edited by David H. Boal and Richard M. Woloshyn

Volume 105—Laser Applications in Chemistry
edited by K. L. Kompa and J. Wanner

Volume 106—Multicritical Phenomena
edited by R. Pynn and A. Skjeltorp

Volume 107—Positron Scattering in Gases
edited by John W. Humberston and M. R. C. McDowell

Volume 108—Polarons and Excitons in Polar Semiconductors and Ionic Crystals
edited by J. T. Devreese and F. Peeters

Series B: Physics

Positron Scattering in Gases

Edited by

John W. Humberston

University College London
London, England

and

M. R. C. McDowell

Royal Holloway College
Egham, Surrey, England

Plenum Press
New York and London
Published in cooperation with NATO Scientific Affairs Division

Proceedings of the NATO Advanced Research Workshop on
Positron Scattering in Gases,
held July 19–23, 1983,
at Royal Holloway College, Egham, Surrey, England

Library of Congress Cataloging in Publication Data

NATO Advanced Research Workshop on Positron Scattering in Gases (2nd:1983:
Royal Holloway College)

Positron scattering in gases.

(NATO ASI series. Series B, Physics; v. 107)
"Proceedings of the NATO Advanced Research Workshop on Positron Scatter-
ing in Gases, held July 19–23, 1983, at Royal Holloway College, Egham, Surrey,
England"—T.p. verso.
"Published in cooperation with NATO Scientific Affairs Division."
Includes bibliographical references and index.
1. Positrons—Scattering—Congresses. I. Humberston, John W. II. McDowell,
M. R. C. III. Title. IV. Series.
QC793.5.P628N38   1983   539.7′214   84-1907
ISBN-13: 978-1-4612-9804-5   e-ISBN-13: 978-1-4613-2751-6
DOI: 10.1007/978-1-4613-2751-6

©1984 Plenum Press, New York
Softcover reprint of the hardcover 1st edition 1984
A Division of Plenum Publishing Corporation
233 Spring Street, New York, N.Y. 10013

This book is dedicated to the memory of
Sir Harrie Massey,
who died on 27th November, 1983.
No worthier testimony to his courage and
his interest in positron physics could be found
than that, despite being seriously ill at the time,
he gave the opening address at the NATO Advanced Research Workshop on
Positron Scattering in Gases.

ORGANIZING COMMITTEE

Margaret F. Dixon	Royal Holloway College
T.C. Griffith	University College London
J.W. Humberston	University College London
M.R.C. McDowell	Royal Holloway College

INVITED SPEAKERS

K.F. Canter	Brandeis University, U.S.A.
M. Charlton	University College London, U.K.
P.G. Coleman	University of Texas at Arlington, U.S.A.
R.J. Drachman	Goddard Space Flight Center, U.S.A.
D.W. Gidley	Bell Laboratories, U.S.A.
T.C. Griffith	University College London, U.K.
G.R. Heyland	University College London, U.K.
R.H. Howell	Lawrence Livermore National Laboratory, U.S.A.
L.D. Hulett, Jr.	Oak Ridge National Laboratory, U.S.A.
F.M. Jacobsen	Riso, Denmark
C.J. Joachain	Université Libre de Bruxelles, Belgium
W.E. Kauppila	Wayne State University, U.S.A.
K.G. Lynn	Brookhaven National Laboratory, U.S.A.
R.P. McEachran	York University, Canada
Sir Harrie Massey	University College London, U.K.
A.P. Mills, Jr.	Bell Laboratories, U.S.A.
R.M. Nieminen	University of Jyväskylä, Finland
W. Raith	Universität Bielefeld, W. Germany
D.M. Schrader	Marquette University, U.S.A.
G. Sinapius	Universität Bielefeld, W. Germany

PREFACE

The first conference in this series, devoted principally to the interaction of positrons in gases, was held at York University, Toronto, in July 1981 immediately preceding the XII ICPEAC in Gatlinburg, and the proceedings were published in the Canadian Journal of Physics, volume 60 (1982). So successful was this meeting that the decision was taken to hold a second one around the time of XIII ICPEAC in Berlin in 1983. London was clearly a convenient location but, rather than the obvious choice of University College London in central London, the Organising Committee decided that the beautiful and peaceful surroundings of Royal Holloway College would provide a more pleasant and intimate atmosphere for a small meeting.

Even a small conference requires substantial sums of money to pay the expenses of invited speakers and when considering possible sources of funds the Organising Committee recognised that the intended format of the meeting and the international composition of the participants made it appropriate to apply to the NATO Science Committee for support under the Advanced Research Workshop Programme. This was one of the few successful applications made this year, and so it was that the conference became the 'NATO Advanced Research Workshop on Positron Scattering in Gases'.

The Workshop, with approximately sixty participants, started after lunch on 19 July, 1983 and finished at mid-day on 23 July. Each day was divided into a morning and afternoon session, and the organisation of each session was the responsibility of two or three of the invited speakers. In addition to their own talks they also included relevant contributions from other participants. Several contributed papers were submitted to the Workshop and these were presented in a poster session.

Only the papers of the invited speakers are included in these proceedings, as it is assumed that the contributed papers will ultimately be published in the usual journals.

Approximately half the sessions were devoted to reviews of the current state of theoretical calculations and experimental results

obtained using conventional positron sources. The remainder of the
time was spent discussing various methods of producing much more
intense positron beams and speculating on some of the experiments
which might be done when these beams become available.

The opening address was given by Sir Harrie Massey who, in
addition to being one of the great pioneers, has maintained a
considerable interest in the field over many years. Most regrettably,
ill health has prevented him from preparing a written version of his
talk for inclusion in these proceedings.

The NATO Science Committee has already been mentioned as the
principal source of funds for the Workshop. The Organising Committee
also gratefully acknowledge contributions from EG and G Instruments
Ltd. and Vacuum Generators Ltd., who both provided exhibitions of
scientific equipment and Royal Holloway College and University
College London. An interest-free loan from the Royal Society is
also gratefully acknowledged.

Several people at Royal Holloway College helped with the
organisation of the Workshop, but a particular debt of gratitude
is owed to Mrs. Margaret Dixon who was the secretary to the Organising Committee until she became ill shortly before the start of
the Workshop. Her place was most ably taken at short notice by
Mrs. Betty Alderman, whose unfailing cheerfulness and efficiency
throughout the meeting was greatly appreciated by all participants.
Mention must also be made of the help given by Mrs. Betty Diggens,
whose first day in her new job coincided with the start of the
Workshop.

It was generally agreed that the Workshop had been very useful
and that regular meetings should take place to review progress in
the field of positron collisions in gases. Already preliminary
plans are being made for a further meeting around the time of the
next ICPEAC in 1985.

                                        J.W. Humberston

                                        M.R.C. McDowell

                                        25th October, 1983

CONTENTS

Survey of Recent Experimental Results on Positron
    Scattering in Gases. Part I: Total Cross
    Sections. 1
  W. Raith

Survey of Recent Experimental Results on Positron
    Scattering in Gases. Part II: Beyond
    Total Cross Sections. 15
  W.E. Kauppila and T.S. Stein

Survey of Recent Theoretical Results on Positron
    Scattering in Gases: Low Energy 27
  R.P. McEachran

Survey of Recent Theoretical Results on Positron
    Scattering in Gases: Intermediate and High
    Energies. 39
  C.J. Joachain

Positronium Formation Cross-Sections in Various
    Gases 53
  T.C. Griffith

Positronium: Recent Fundamental and Applied Research 65
  D.W. Gidley and P.G. Coleman

Positronium Formation in Gases and Liquids. 85
  F.M. Jacobsen

Positron Lifetime Spectra for Gases 99
  G.R. Heyland

The Calculation of Positron Lifetimes 109
  D.M. Schrader

Techniques for Studying Systems Containing Many
    Positrons 121
  A.P. Mills, Jr.

Surface Studies with Slow Positron Beams. 139
    R.M. Nieminen

Intense Positron Beams: Linacs. 155
    R.H. Howell, R.A. Alvarez, K.A. Woodle, S. Dhawan,
      P.O. Egan, V.W. Hughes and M.W. Ritter

Intense Positron Beams and Possible Experiments 165
    K.G. Lynn and W.E. Frieze

Intense Postiron Beams - An Evaluation of the Methods 181
    M. Charlton

The Generation of Monoenergetic Positrons 195
    L.D. Hulett, Jr., J.M. Dale, P.D. Miller, Jr.,
      C.D. Moak, S. Pendyala, W. Triftshäuser, R.H.
      Howell, R.A. Alvarez

Applications of Intense Positron Beams. 203
    R.J. Drachman

Application of Intense Positron Beams in Atomic Physics
      Experiments 211
    G. Sinapius

Low Energy Positron and Positronium Diffraction 219
    K.F. Canter

Index 227

SURVEY OF RECENT EXPERIMENTAL RESULTS ON POSITRON SCATTERING

IN GASES PART I: TOTAL CROSS SECTIONS

Wilhelm Raith

Universität Bielefeld
Fakultät für Physik
D-4800 Bielefeld 1, Fed. Rep. of Germany

Introduction

This survey concentrates on recent developments not covered in the invited paper of Kauppila and Stein (1982), given at the 1981 Toronto Conference, and the recent review articles of Griffith (1979) and Griffith and Heyland (1978). Updated information on this topic can be found in the reports of these authors presented at the Arlington Conference last year (Stein and Kauppila, 1982; Griffith et al., 1982). Electron total cross section measurements were thoroughly reviewed by Bederson and Kieffer in 1971. Their discussion of systematic errors is still up-to-date and also pertinent to measurements with positrons. The advantages and limitations of time-of-flight (TOF) methods in scattering spectroscopy were reviewed by Raith (1976).

With regard to the total cross section measurements made thus far, the positron studies correspond to the electron studies of the late 1920's when a variety of non-corrosive room-temperature target gases had been investigated and experiments with alkali metal vapors were being prepared. With regard to experimental sophistication and accuracy, however, the state of art of recent positron experiments is quite comparable to modern electron experiments. Many positron total cross section measurements are combined with electron measurements in the same apparatus in order to obtain the cross-section ratio

$$R(E) = \sigma^+(E) \ / \ \sigma^-(E)$$

with greatly reduced systematic errors. The attention paid to R(E) has also led to a renewed interest in electron total cross sections.

Because of new technologies for the production of high-intensity beams of moderated positrons, partial and differential cross section measurements will soon be feasible. Should all efforts be directed toward the more sophisticated experiments or is it also worthwhile to expand and improve the total cross section measurements?

Techniques

The total cross section measurements to be discussed here are all transmission-type experiments based ideally on the equation

$$I(E) = I_o(E) \exp [- \rho \ell \sigma(E)]$$

where I and I_o are the transmitted intensities with and without gas in the target cell, E is the positron energy, ρ the number of target molecules per unit volume, ℓ the target length, and σ the total cross section.

In the early experiments the positrons were detected by means of 2γ-coincidences, utilizing the unique positron signature provided by the two-photon annihilation for background discrimination. Later the channel electron multiplier became the favored detector because of its high counting efficiency for charged particles and the advantage of using one detector for both positrons and electrons.

In most positron transmission experiments the slow positrons are deflected by a large angle in order to shield the detector against γ-radiation and high-energy positrons from the source. For intensity reasons a longitudinal magnetic guiding field is employed in some spectrometers. Deflection of the slow positrons in the presence of a longitudinal magnetic field is accomplished by a curved solenoid. Only with the channel electron-multiplier detector (which is nearly γ-blind) and a TOF method (which allows easy discrimination of the "prompt" high-energy positrons) is a straight-line geometry possible.

TOF methods played an important role in positron total cross section measurements from the beginning. The first experiment (Costello et al. 1972) was performed with a pulsed electron accelerator for which TOF was an obvious choice. Shortly thereafter the London group (Coleman et al., 1972) introduced a TOF method for radioactive positron sources with single-particle zero-time pickoff where each positron generates its own 'start' time mark by penetrating a thin plastic scintillator before being moderated. In TOF a broad energy distribution does not limit the energy resolution because for every particle the energy is determined from the measured flight time. But a disadvantage of this positron TOF method is its inherent intensity limitation. The event rate

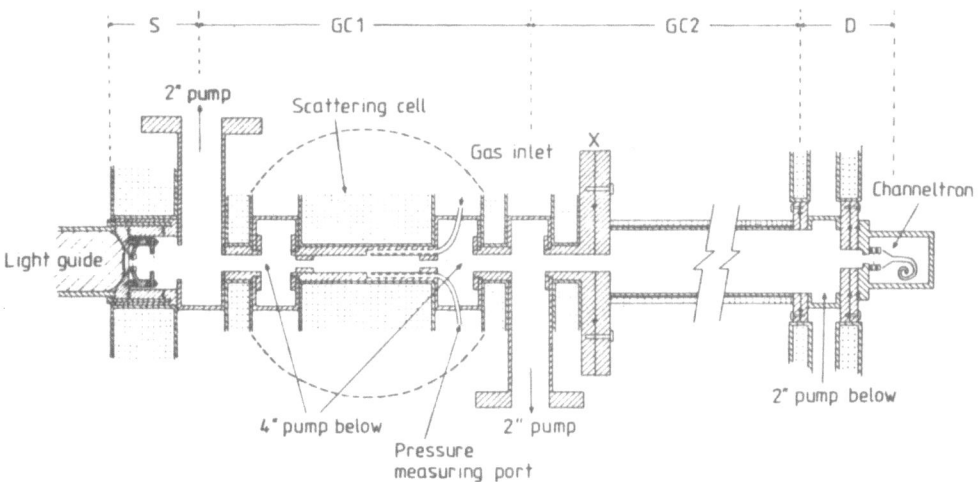

*Fig. 1.   Schematic diagram of the apparatus used by the London
group (from Charlton et al., 1983). S: source region,
GC1: gas cell region of 0.4 m length, GC2: auxilliary
section of flight tube which extends the system length
to 0.77 m when inserted, D: detector region. (Drawing
not to scale) -- Technical details not shown: Na-22
source (100 μCi) / venetian blind W-moderator / TOF with
inverted timing / variation of acceleration voltage
between moderator and grid / longitudinal magnetic field
with field-strength ratios of 6:1:8 for source, gas cell
and detector regions, respectively/positrons scattered
through large angles are reflected in the inhomogeneous
magnetic field before reaching the detector / flight-
time discrimination is employed to suppress positrons
scattered through smaller angles / the gas cell is
80 mm long with 6 mm apertures at each end / the gas
density integral along the axis was determined by
normalization to well-known cross sections.*

of the detector connected to the scintillator is orders of magnitude
higher than that of the detector at the end of the flight path. The
necessity of correlating start and stop pulses unambiguously limits
the source activity to about 100 μCi. At present, only the London
group uses a TOF spectrometer for total cross section measurements
(Fig.1). An improved version of the Bielefeld TOF positron spectrometer (Sinapius et al., 1980) designed for total cross section
measurements below 10 eV is being built.

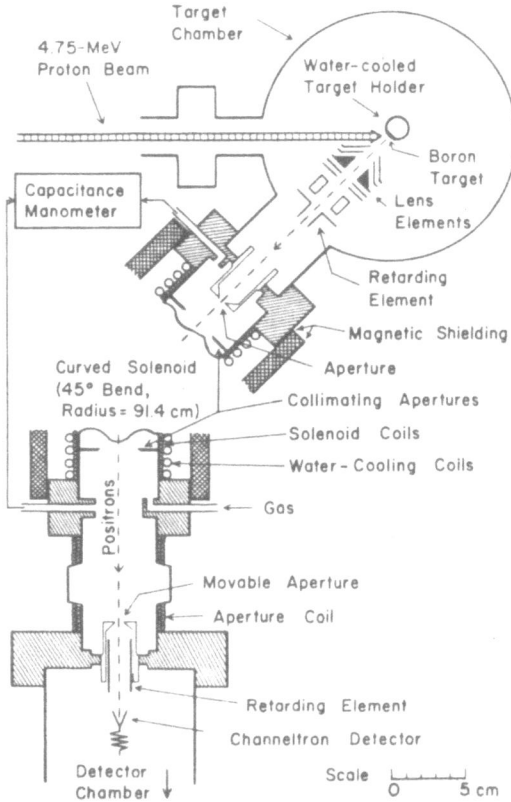

Fig. 2. Schematic diagram of the apparatus used by the Detroit group (from Kauppila et al., 1981) -- Technical details not shown: Radioactive source produced by $^{11}B(p,n)^{11}C$ reaction / boron also serves as moderator / moderated positrons were found to have very small energy spread of about 0.1 eV / for electron measurements a thermionic cathode is inserted / retarding element is employed to discriminate against electrons or positrons which lost energy or were scattered through larger angles.

Of all the systematic errors involved in positron total cross section measurements the discrimination against forward scattered positrons was discussed most thoroughly (Kauppila and Stein, 1982). Apparently this error is now understood for every spectrometer. Systematic errors in target-gas manometry are small; the capacitance manometer (Baratron) is used by everybody. The effective length of the target cell is more difficult to determine. If the whole space between moderator and detector is filled with the target gas,

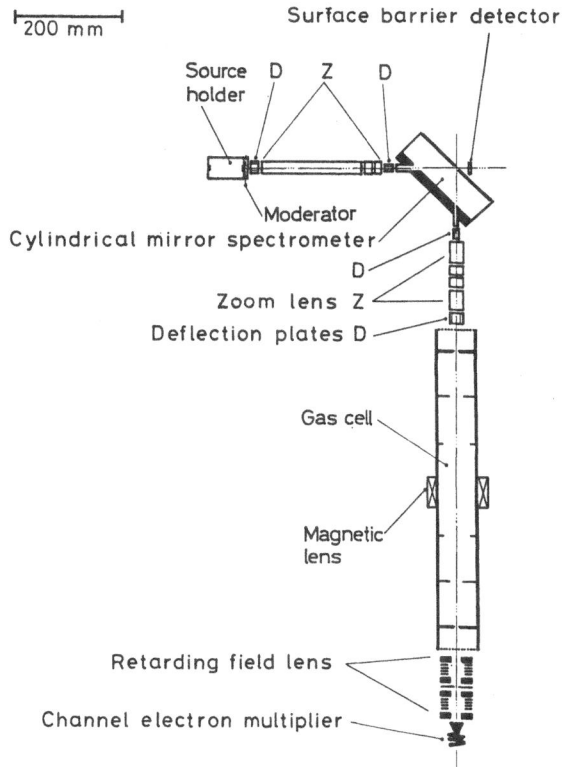

Fig. 3. Schematic diagram of the apparatus used by the Bielefeld group -- Technical details not shown: Na-22 source (3 mCi) / W-vane moderator / electrostatic 90° beam bending / gas flow is kept constant by bypassing the gas cell in the "without gas" mode / electron measurements are made with secondary electrons from the moderator.

corrections to the measured cross section values are necessary in order to account for scattering in regions where the positrons have different energy. If a localized target cell is used with differential pumping at entrance and exit, molecular beams effuse from both orifices. The density integral along the beam axis can be determined by calibrating the target cell via the measurement of a well-known cross section. It is also possible to calculate the density distribution of effusive-flow patterns (Mathur and Colgate, 1972; Nelson and Colgate, 1973; Mathur et al., 1975).

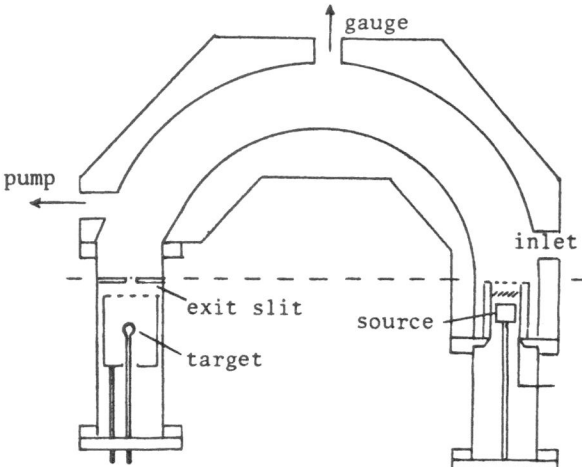

*Fig. 4. Schematic diagram of the apparatus used by the Swansea
         group (from Dutton et al., 1982) -- Technical details not
         shown: 180° bend on 15 cm radius in transverse magnetic
         field / resolvable energy ΔE is about 5% of energy E /
         Na-22 source (3 mCi) / venetian blind moderator with MgO
         covered Au-vanes / acceleration of positrons between
         moderator and grid / deceleration of positrons down to
         10 eV before annihilation on the target / detection by
         2γ-coincidence / the angular discrimination by means of
         retarding potential lies between 5.7° and 3.3°.*

Spectrometers in use

According to what has been published after the Toronto
Conference only four positron spectrometers are currently in use
for measuring total cross sections. These four distinctly different
set-ups are shown here (Fig.1-4) in order to point out that their
systematic errors will most likely be distinctly different too.
Therefore, consistency of results can be interpreted as an indication of accuracy which is more convincing than the estimates of
systematic errors given by the authors.

The London group has employed the positron TOF method since
1972. They also developed an important TOF data-evaluation procedure (Coleman et al., 1974 b). Several improvements in spectrometer design led to the version shown in Fig.1 which has been used
since 1977. The other three spectrometers (Fig.2-4) do not utilize
TOF. In those spectrometers the resolvable energy ΔE is determined
by the energy spread of the moderated positrons. The positron beam
intensity depends on moderator efficiency and source activity. The
latter is much higher than that permitted in a TOF spectrometer.

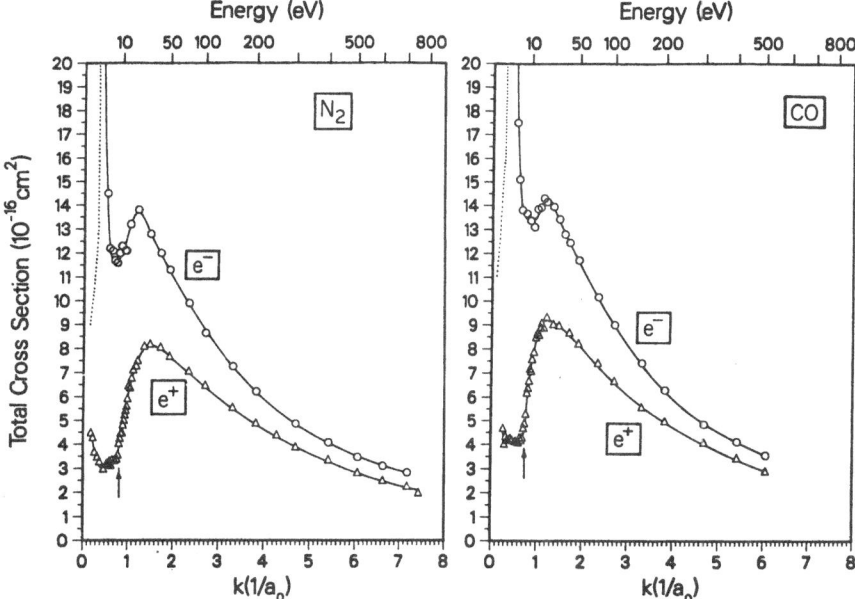

Fig. 5.   *Total $e^{\pm}$ cross section measurements on the isoelectronic target molecules N_2 and CO (adopted from Hoffman et al., 1982, and Kwan et al., 1983). The similarity of the cross-section curves for isoelectronic molecules was also demonstrated for N_2O (Kauppila et al., 1983) and CO_2 (Kwan et al., 1983)*

The Detroit group invented the method based on boron activation by proton bombardment (Fig.2) and has used it most successfully in numerous measurements, performed on-line of their own proton accelerator. The Bielefeld spectrometer (Fig.3) designed for measurements at energies above 10 eV was used both with a weak Na-22 source at the home laboratory and a strong C-11 source produced on-line of the proton accelerator at Bochum (Deuring et al., 1983). Because of the problems associated with the $^{11}B(p,n)^{11}C$ method and with the logistics of guest work at a distant accelerator it is now being used with Na-22 only. It is planned to switch from 3 to 50 mCi and to use the electrostatic 90° deflection at a low beam energy for reducing the energy spread of positrons and electrons.

The Swansea spectrometer (Fig.4) permits positron measurements up to 3 keV which is a significant extension of the energy range covered by the other groups, important for investigating the region where R(E) approaches unity.

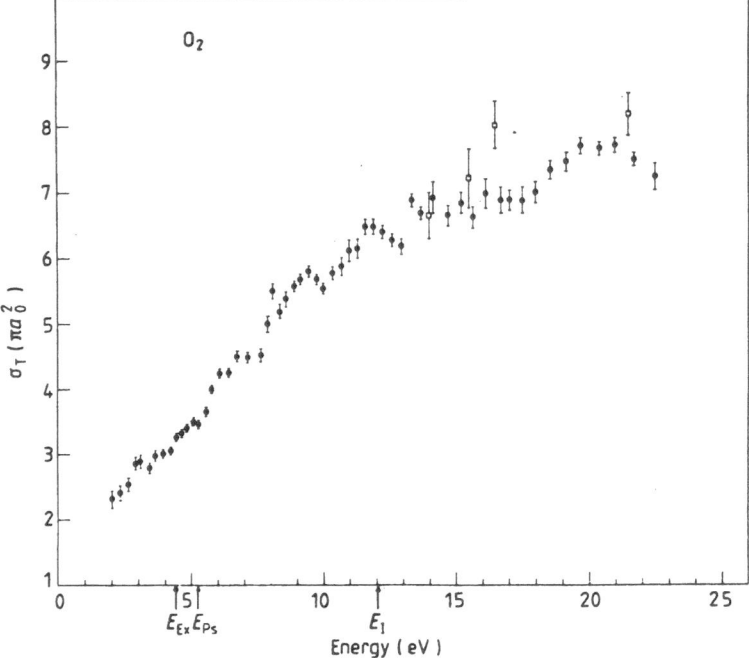

*Fig. 6. e^+-O_2 total cross sections (from Charlton et al., 1983).
The open squares are data points measured earlier
(Charlton et al., 1980).*

Data published recently

The detroit group published $e^{\pm}$ total cross sections for H_2,
N_2, and CO_2 (Hoffman et al., 1982), Kr and Xe (Dababneh et al.
(1982), CO and CO_2 (Kwan et al., 1983). All the Detroit data are
distinguished by the high energy resolution over a large energy
range extending from 0.5 to 800 eV. As an example the data for
N_2 and CO are shown in Fig.5. The assumption that $R = \sigma^+/\sigma^-$ is
generally smaller than unity (except for the region of the electron
Ramsauer minima of Ar, Kr, Xe) and approaches unity at high energies is not true for all targets. For H_2 and also Xe measurements
show that $R(E)$ is greater than unity at intermediate energies.
This anomalous behavior is attributed to the positronium formation
channel for which no equivalent exists in electron inelastic
scattering. The London group published positron cross sections
for H_2, N_2, CO_2, O_2, and CH_4 (Charlton et al., 1983) in the energy
range of 2 to 20 eV. The TOF method provides good energy resolution, sufficient to resolve steps in $\sigma^+(E)$ for H_2 at the thresholds
of positronium formation, first electronic excitation and ionization. The e^+-O_2 data shown in Fig.6 do not exhibit steps at

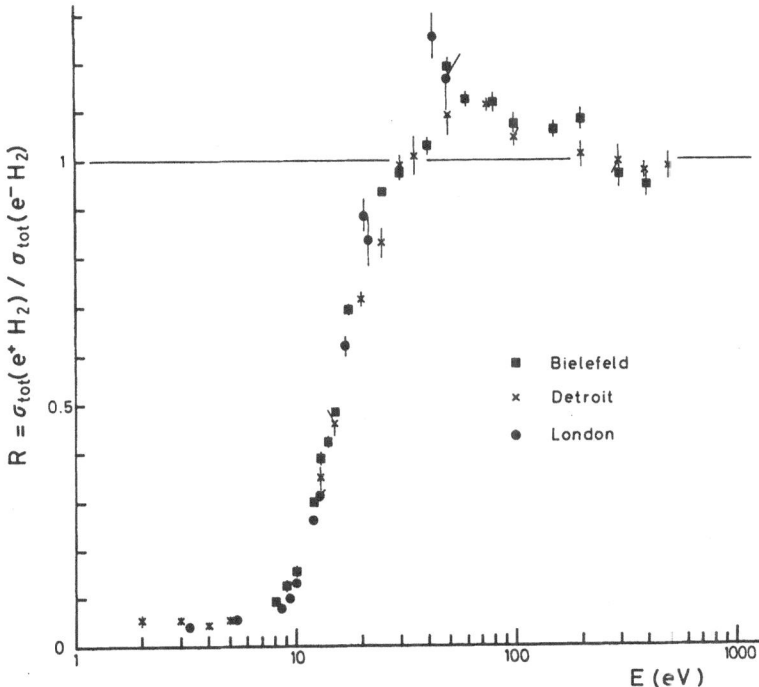

Fig. 7.   The ratio R of positron-to-electron total cross section
           vs. energy for molecular hydrogen according to published
           data from Bielefeld (Deuring et al., 1983), Detroit
           (Hoffman et al., 1982) and London (Charlton et al., 1980;
           Griffith et al., 1982).

those thresholds but other rather intriguing structure of yet unknown origin. The Swansea group reported e^+-N_2 cross sections
between 20 and 3000 eV (Dutton et al., 1982) in good agreement with
London and Detroit data in the overlap region at lower energy.

    As the first target gas to be studied in our new spectrometer
(Fig.3) we chose H_2 in order to facilitate a critical comparison
with the results from London and Detroit. Our $e^{\pm}$-H_2 data cover
the energy range between 8 and 400 eV. The general agreement with
the other two groups is good, except for σ^+ between 30 and 70 eV
where we agree well with Detroit while the London cross sections
(Charlton et al., 1980) are somewhat higher. In the range of the
inelastic thresholds, where $\sigma^+(E)$ rises steeply, the data of the

three groups differ from each other in a systematic way suggesting
differences in energy scale calibration. The electron data of the
three groups are in good agreement. All three groups found
R(E) > 1 at intermediate energies (Fig.7).

What can be expected in the near future ?

There are still some interesting non-corrosive room-temperature
gases for which positron cross sections should be measured (e.g.,
SF_6, C_2H_6, C_2H_4, C_2H_2 ...) Current work at Detroit (Kauppila et al.,
1983) and Bielefeld (Floeder et al., 1983) reported at this Workshop
is going in this direction. The alkali metals require target cells
at moderately high temperatures. As long as positron measurements
with atomic hydrogen are not feasible, the alkali metals are the
only one-electron atoms accessible. A special feature of the
alkali metals is that they have smaller ionization energies than
positronium, thus there is no Ps-formation threshold. The Detroit
group reported at this Workshop the first e^+-K total cross section
measurements (Stein et al., 1983) which is a great experimental
accomplishment. The development of high-temperature targets will
probably lead to measurements with other metals (Hg, Cd, Tl, Zn ...)
in analogy to the development of the electron measurements (Brode,
1933). Because low-energy electron scattering from polar molecules
(H_2O, CsF ...) is a rather difficult problem theoretically (Garret,
1972), electron and positron cross section measurements are wanted.

For all targets there are three points of particular interest
in total cross section measurements:
(1) Search and study of pronounced structures (thresholds, reso-
     nances, cusps ...) with high energy resolution
(2) Positron and electron measurements in the convergence region
     where R(E) → 1
(3) Measurements at very low energies

Fig.6 shows an example of structure which demands further in-
vestigation. Although it is true that partial cross section measurements for individual reaction channels can provide much more
insight in the physics involved, it is also a fact that partial
and differential measurements are much more difficult and, therefore, will not be competitive with total cross section measurements in accuracy and energy resolution. Since the convergence
region is of great theoretical interest, the energy range of $e^\pm$
measurements should be extended in order to reach this region for
all targets. With Ps-formation cross sections becoming available,
as reported by Griffith (London) and Coleman (Arlington) at this
Workshop, it will soon be possible to determine the ratio

$$R_{no\,Ps}(E) = [\sigma^+(E) - \sigma_{Ps}(E)] / \sigma^-(E)$$

which should not exhibit the anomalous behavior of $R(E)$ seen in H_2 (Fig.7) and Xe (Dababneh et al., 1982) and be a better indicator of Born-approximation validity.

At very low energies the total cross sections are easier to interpret than at higher energies because both the number of open channels and the number of significantly contributing partial waves decrease with energy. The "modified effective range theory" (MERT) provides guidance for fitting $\sigma(E)$ data to a theoretically plausible equation with only a few free parameters which can be extrapolated to zero energy, where $\sigma = 4\pi A^2$ and A is the scattering length (O'Malley, 1963). MERT, originally derived for isotropic long-range potentials and applied only to noble-gas atom cross sections, was recently extended to electron scattering from molecules by Chang (1981). Improved schemes for data fitting were suggested by O'Malley and Crompton (1980) and Paul (1980). From an experimental point of view MERT is an essential tool for critical comparison of low-energy total cross sections with the momentum-transfer cross sections derived from drift-velocity measurements (Huxley and Crompton, 1974). At $E = 0$ both cross sections are equal. The first positron drift measurement was performed by Böse et al. (1981) in molecular hydrogen. These authors conclude from their data that the e^+-H_2 cross section at zero energy is at least four time the e^--H_2 cross section. The lowest-energy data points available for H_2 show only a steep rise of $\sigma^+(E)$ with decreasing energy without any indication of leveling off. The situation is similar for other molecular gases as well as all the noble gases. Measuring $\sigma^+(E)$ for $E \rightarrow 0$ is a challenge.

REFERENCES

Bederson, B. and Kieffer, L. J., 1971, Total electron-atom collision cross sections at low energies – a critical review, Rev. Mod. Phys., 43:601.

Böse, N., Paul, D. A. L., and Tsai, J.-S., 1981, Positron drift in molecular hydrogen, J. Phys. B, 14:L227.

Brode, R. B., 1933, The quantitative study of the collisions of electrons with atoms, Rev. Mod. Phys., 5:257.

Chang, E. S., 1981, Modified effective-range theory for electron scattering from molecules, J. Phys. B, 14:893.

Charlton, M., Griffith, T. C., Heyland, G. R., and Wright, G. L., 1980, Total scattering cross sections for intermediate-energy positrons in the molecular gases H_2, O_2, N_2, CO_2, and CH_4, J. Phys. B, 13:L353.

Charlton, M., Griffith, T. C., Heyland, G. R., and Wright, G. L., 1983, Total scattering cross section for low-energy positrons in the molecular gases H_2, N_2, CO_2, O_2 and CH_4, J. Phys. B, 16:323.

Coleman, P. G., Griffith, T. C., and Heyland, G. R., 1972,
    A method of improving the statistical accuracy in
    lifetime measurements on positrons annihilating in gases,
    J. Phys. E, 5:376.
Coleman, P. G., Griffith, T. C., and Heyland, G. R., 1974 a,
    Measurement of total scattering cross sections for
    positrons at energies 2-400 eV on molecular gases:
    H_2, D_2, N_2, CO, Appl. Phys., 4:89.
Coleman, P. G., Griffith, T. C., and Heyland, G. R., 1974 b,
    The analysis of data obtained with time to amplitude
    converter and multichannel analyser systems, Appl. Phys.
    5:230.
Costello, D. G., Groce, D. E., Herring, D. F., and McGowan, J. W.,
    1972, (e^+, He) total scattering, Can. J. Phys., 50:23.
Dababneh, M. S., Hsieh, Y.-F., Kauppila, W. E., Pol, V., and
    Stein, T. S., 1982, Total-scattering cross-section
    measurements for intermediate energy positrons and
    electrons colliding with Kr and Xe, Phys. Rev. A,
    26:1252.
Deuring, A., Floeder, K., Fromme, D., Raith, W., Schwab, A.,
    Sinapius, G., Zitzewitz, P. W., and Krug, J., 1983,
    Total cross section measurements for positron and
    electron scattering on molecular hydrogen between 8 and
    400 eV, J. Phys. B, 16:1633.
Dutton, J., Evans, C. J., and Mansour, H. L., 1982, Total cross-
    sections for positron scattering in nitrogen at energies
    from 20 to 3000 eV, in: "Positron Annihilation",
    P. G. Coleman, S. C. Sharma, and L. M. Diana, eds.,
    North-Holland, Amsterdam.
Floeder, K., Fromme, D., Raith, W., Schwab, A., and Sinapius, G,
    1983, The Bielefeld e^+/e^- total scattering experiment,
    Contribution to this Workshop.
Garret, W., 1972, Low-energy electron scattering by polar molecules,
    Molec. Phys., 24:465.
Griffith, T. C. and Heyland, G.R., 1978, Experimental aspects of
    the study of the interaction of low-energy positrons with
    gases, Phys. Rev., 39:169.
Griffith, T. C., 1979, Experimental aspects of positron collisions
    in gases, Adv. At. Molec. Phys., 15:135.
Griffith, T. C., Charlton, M., Clark, G., Heyland, G. R., and
    Wright, G. L., 1982, Positrons in gases - a progress
    report, in: "Positron Annihilation", P. G. Coleman,
    S. C. Sharma and L. M. Diana, eds., North-Holland,
    Amsterdam.
Huxley, L. G. H. and Crompton, R. W., 1974, "The diffusion and
    drift of electrons in gases", John Wiley & Sons,
    New York.

Kauppila, W. E., Stein, T. S., Smart, J. H., Dababneh, M. S., Ho, Y. K., Downing, J. P., and Pol, V., 1981, Measurements of total scattering cross sections for intermediate-energy positrons and electrons colliding with helium, neon, and argon, Phys. Rev. A, 24:725.

Kauppila, W. E. and Stein, T. S., 1982, Positron-gas cross-section measurements, Can. J. Phys., 60:471.

Kauppila, W. E., Dababneh, M. S., Hsieh, Y.-F., Kwan, Ch. K., Smith, S., Stein, T. S., and Uddin, M. N., 1983, $e^{+,-}$-N_2O, CH_4, and SF_6 total scattering measurements, Contribution to this Workshop.

Kwan, C. K., Hsieh, Y.-F., Kauppila, W. E., Smith, S. J., Stein, T. S., and Uddin, M. N., 1983, $e^{\pm}$-CO and $e^{\pm}$-CO_2 total cross-section measurements, Phys. Rev. A, 27:1328.

Mathur, B. P. and Colgate, S. O., 1972, Calculations of effusive-flow patterns. I. Knudsen-cell results, Phys. Rev. A, 6:1266.

Mathur, B. P., Field, J. E., and Colgate, S. O., 1975, Calculations of effusive-flow patterns. III. scattering chambers with thin circular apertures, Phys. Rev. A, 11:830.

Nelson, R. N. and Colgate, S. O., 1973, Calculations of effusive-flow patterns. II. scattering chambers with semi-infinite slits, Phys. Rev. A, 8:3045.

O'Malley, T. R., 1963, Extrapolation of electron-rare gas atom cross sections to zero energy, Phys. Rev., 130:1020.

O'Malley, T. F. and Crompton, R. W., 1980, Electron-neon scattering length and S-wave phaseshifts from drift velocities, J. Phys. B, 13:3451.

Paul,D., 1980, A speculation about effective range formulas applicable to S-wave scattering under long range polarization potentials, Can. J. Phys., 58:134.

Raith, W., 1976, Time-of-flight scattering spectroscopy, Adv. At. Molec. Phys., 12:281.

Ramsauer, C. and Kollath, R., 1930, Über den Wirkungsquerschnitt der Nichtedelgasmoleküle gegenüber Elektronen unterhalb 1 Volt, Ann. Phys. (Leipzig). 4:91.

Sinapius, G., Raith, W., and Wilson W. G., 1980, Scattering of low-energy positrons from noble-gas atoms, J. Phys. B, 13:4079.

Stein, T. S. and Kauppila, W. E., 1982, Positron-gas collision experiments, in: "Positron Annihilation", P. G. Coleman, S. C. Sharma and L. M. Diana, eds., North-Holland, Amsterdam.

Stein, T. S., Gomez, R. D., Kwan, Ch.-K., Hsieh, Y.-F., and Kauppila, W. E., 1983, $e^{+,-}$-K total scattering cross section measurements, Contribution to this Workshop.

SURVEY OF RECENT EXPERIMENTAL RESULTS ON POSITRON SCATTERING IN GASES

PART II: BEYOND TOTAL CROSS SECTIONS

W.E. Kauppila and T.S. Stein

Department of Physics and Astronomy
Wayne State University
Detroit, Michigan 48202 USA

INTRODUCTION

The first decade (1972-82) of positron-gas scattering experiments performed with slow positron beams has been primarily concerned with the measurements of total scattering cross sections for positrons colliding with the inert gases and various molecules, which was the subject of the preceding paper presented at this Workshop.[1] In this paper the focus of attention will be placed on positron scattering experiments performed with slow positron beams which extend beyond total cross section measurements. Included in this category are measurements of differential and inelastic cross sections for the scattering of positrons by atoms and molecules, and searches for positron scattering resonances.

DIFFERENTIAL SCATTERING

It is well known that measurements of differential scattering cross sections can provide a more sensitive test of scattering theories than measurements of total scattering cross sections because more information is contained in differential cross sections relating to the various individual scattering phase shifts. The first reported differential scattering measurements have been made by Coleman and McNutt,[2] where they have determined the differential cross sections for the elastic scattering of 2-9 eV positrons by Ar for angles ranging from 20-60°. A schematic diagram of their apparatus,[3] which employs a time-of-flight (TOF) approach, is shown in Fig. 1. In their experiment slow positrons pass through a 1 cm long gas cell, travel 25 cm in an evacuated straight flight tube (possessing a strong axial magnetic field), and then are detected by

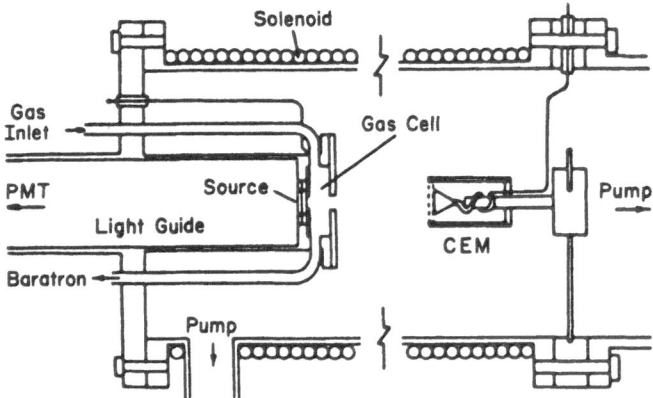

Fig. 1. Texas TOF spectrometer. (From Coleman et al., Ref. 3)

a channeltron electron multiplier (CEM). By alternately admitting
gas into and then evacuating the gas cell TOF histograms were
obtained as shown in Fig. 2. The TOF spectrum obtained when gas is
present in the scattering cell is found to have a "tail" on the
long-time side which relates to detected positrons that have
experienced forward elastic scattering. An appropriately adjusted
"vacuum" TOF spectrum is then subtracted from the "gas" spectrum in
order to obtain a "difference" spectrum. This "difference" spectrum
is then correlated to various angles of forward elastic scattering
in order to obtain differential cross sections. The results of
their measurements are compared with the calculations of Schrader[4]
(solid lines) and "scaled-down" calculations of McEachran et al.[5]

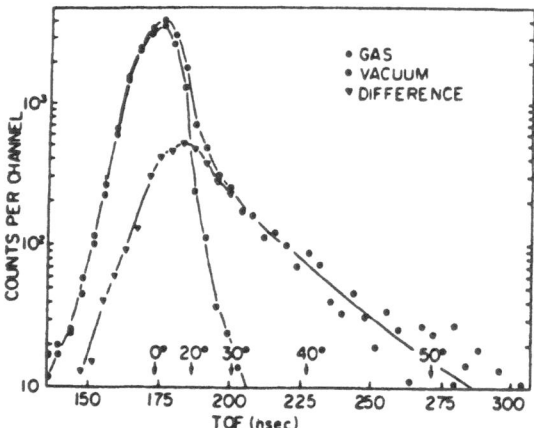

Fig. 2. TOF histogram for e^+-Ar scattering at 6.7 eV for a run
         time of 55,000 s. (From Coleman and McNutt, Ref. 2)

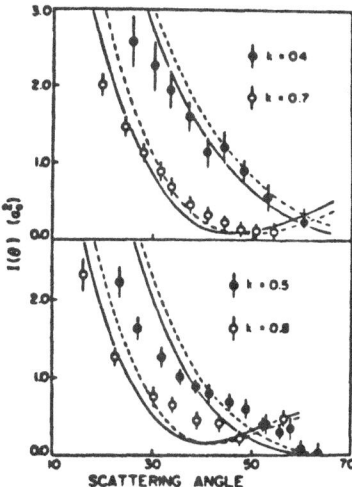

Fig. 3.   Differential e$^+$-Ar scattering measurements compared with
          theory. (From Coleman and McNutt, Ref. 2)

(broken lines) in Fig. 3. The agreement between the experiment and
theory is reasonable. Coleman et al.[3] have also measured
differential elastic scattering cross sections in the angular range
of 25–60° for 5–20 eV electrons colliding with argon in the same
apparatus and using the same technique as used for their positron
studies and obtain reasonable agreement with the measurements of
Williams.[6]

   In our laboratory we have built an experimental system,[7] shown
in Fig. 4, for making direct measurements of positron differential
scattering cross sections. Our approach is to use a crossed-beam
technique in a "field-free" (magnetic fields < 10 mG) scattering

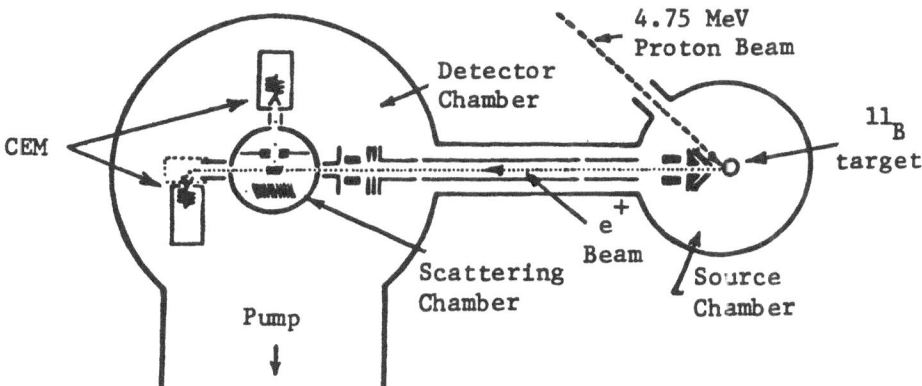

Fig. 4.   Wayne State differential scattering apparatus.

region where the projectile positron beam passes through a target
gas beam (effusing from a multi-channel capillary array source) and
the scattered positrons are to be detected at various angles of
scattering. Since the primary beam and scattered positrons will be
detected by CEMs (which require pressures of less than 10^{-4} torr for
proper operation) it was necessary to incorporate differential
pumping between the scattering region and the region containing the
CEMs. We have tested and refined our system by using an electron
beam and find that we can obtain ratios of detected scattered
electrons to detected primary beam electrons of between 10^{-4} and
10^{-5}, which compares favorably with other electron differential
scattering experiments.[8] Relative differential cross section
measurements that we have made at 90° for 4–20 eV electrons
scattering from He and Ar are in quite good agreement with some
available theories.[9,10] At present our limitation on making
positron differential scattering measurements is having a
sufficiently intense positron beam (>5000/sec).

The Bielefeld group has also built a crossed-beam differential
scattering experiment[11] which they are currently testing with
electrons. Their approach differs from the Detroit experiment in
that they are using cryo-pumping to condense their target gas (Ar
and CO_2) beams which will enable them to use channelplates to detect
the scattered projectile particles at various angles simultaneously
without the need to use differential pumping between the scattering
and detector regions.

The anticipated crossed-beam scattering experiments will have
the distinct advantage of being a direct method for obtaining
differential cross sections, which can be measured over a much
larger range of scattering angles than the TOF approach using a gas
cell. On the other hand more intense positron beams are required
for the crossed-beam approach.

INELASTIC SCATTERING

Some appreciation for the importance of inelastic processes in
positron-gas scattering can be obtained by considering the low
energy positron-inert gas total cross section curves[12–15] displayed
in Fig. 5. In these curves the threshold energies for positronium
formation, atomic excitation, and atomic ionization are indicated,
respectively, by the three arrows in the order of increasing energy.
Below the positronium formation threshold for each of these gases
only elastic scattering can occur. In each case an abrupt increase
in the total scattering cross section is observed after the
positronium formation threshold. If the elastic scattering cross
sections remain smoothly varying as the positron energy is increased
through the positronium formation threshold and the other inelastic
thresholds, it can be seen that the contribution of inelastic

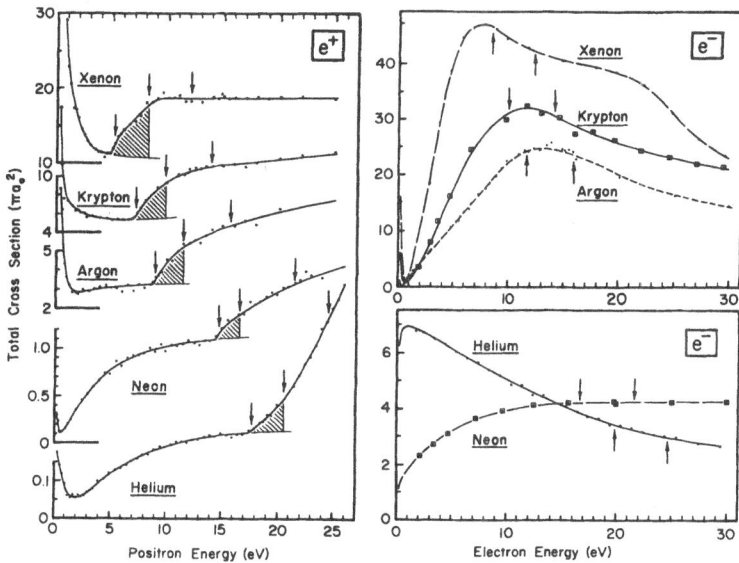

Fig. 5.   Total cross section curves for e$^+$ and e$^-$-inert gas
          scattering. (From Kauppila and Stein, Ref. 12)

processes to the respective total cross section curves at higher
energies becomes comparable to or even greater than the elastic
scattering cross sections. In marked contrast to the positron
curves in Fig. 5 are the corresponding electron-inert gas total
scattering cross section curves[12-15] shown in Fig. 5, which do not
exhibit any noticeable increases in their total cross section curves
as the electron energy is increased through the lowest inelastic
threshold energies for atomic excitation and ionization. It is
interesting that comparisons of total cross section measurements for
positrons and electrons scattered by the inert gases extending up to
intermediate energies[16,17] show that for positron scattering the
maxima of the total cross section curves appear to be associated
with inelastic processes, while for electrons the maxima are related
to elastic scattering.

Positronium Formation

        It is noteworthy that the most dramatic changes in the shapes
of the positron curves shown in Fig. 5 occur at the respective
thresholds for positronium (Ps) formation. From these curves it is
possible to make estimates of the Ps formation cross sections (the
cross-hatched regions) between the thresholds for Ps formation and
atomic excitation, if one assumes that the elastic cross section is
smoothly varying in this energy range. Estimates, such as these,
and observations of the shapes of these curves as the positron
energy increases through the thresholds for atomic excitation and

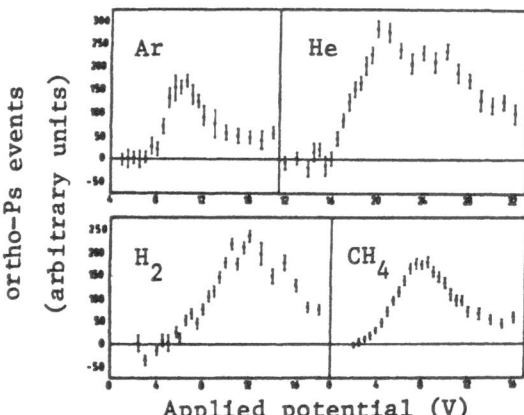

Fig. 6. Energy dependence of ortho-Ps formation. (From Charlton
         et al., Ref. 18)

ionization, provide a strong indication that Ps formation
contributes significantly to the inelastic scattering cross sections
for the inert gases at these low energies.

     The first direct measurements of Ps formation in a beam
scattering experiment have made by Charlton et al.,[18] where they
have measured the energy dependence of the ortho-Ps formation cross
section for He, Ar, H_2, and CH_4, which are shown in Fig. 6. Their
approach was to pass a slow positron beam through a scattering cell
(containing the target gas) and to detect in triple coincidence the
three annihilation gamma rays resulting from the decay of ortho-Ps
which has been produced in the gas scattering cell. In all four
cases ortho-Ps formation energy dependence curves were observed to
reach a maximum within several eV of the threshold energy. It is
interesting that the peaks for helium and argon occur close to the
atomic excitation thresholds for these gases.

     Some preliminary indirect measurements of Ps formation in
helium have been reported by Cook et al.,[19] where the approach was
to pass a slow positron beam through a gas cell and to detect all
the positrons (scattered and unscattered) which did not form Ps.
Their measurements between 9 and 26.5 eV show a noticeable drop in
the collected slow positron beam flux above the Ps formation
threshold (17.8 eV).

     An approach that we have used in our laboratory to make a
preliminary observation of Ps formation in argon[20] is the same as
that of Charlton et al.[18] except that we have detected in coincidence the two annihilation gamma rays resulting from the decay of
para-Ps. We have detected a measurable signal at 10.5 and 12 ev

with the signal at 12 eV being about twice that at 10.5 eV.

A more complete discussion of Ps formation in gases will be provided in a later session of this Workshop.

Excitation and Ionization

From the low energy total cross section curve for positron scattering by helium, shown in Fig. 5, it can be seen that there is a noticeable increase in the slope of the curve after the threshold for atomic excitation, which indicates that inelastic processes other than Ps formation may be contributing significantly to the total cross section. If the Ps formation cross sections for the inert gases reach a maximum within a few eV of their threshold energy, as the relative ortho-Ps formation measurements of Charlton et al.[18] indicate, we have a further indication that atomic and molecular excitation and ionization may contribute appreciably to total positron scattering cross sections.

Using the TOF apparatus shown in Fig. 1 Coleman and Hutton[21] have obtained lower bounds on the total excitation cross sections for 23-31 eV positrons scattering from helium. In this energy range well-defined secondary peaks were observed in the TOF spectra (as shown in Fig. 7 for an incident positron energy of 25.8 eV) which are consistent with positrons losing 20.6 eV of energy and being scattered in the forward direction at angles less than $60°$. The location and shape of the 20.6 eV energy loss peaks are taken as an indication that excitation of the 2^1S state of helium dominates the total excitation cross section in this energy range and that there exists a strong small-angle (less than $20°$) lobe in the angular distribution of the scattered positrons. The partial

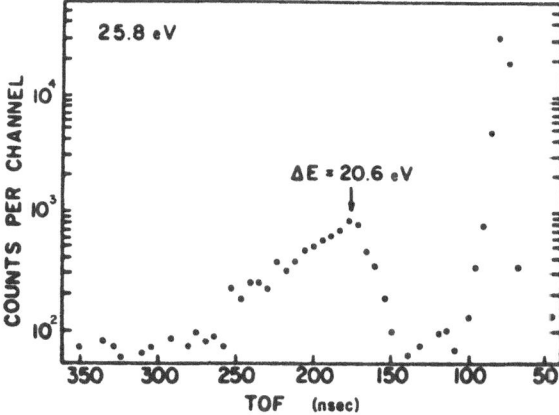

Fig. 7.   TOF spectrum for 25.8 eV positrons colliding with He. (From Coleman and Hutton, Ref. 21)

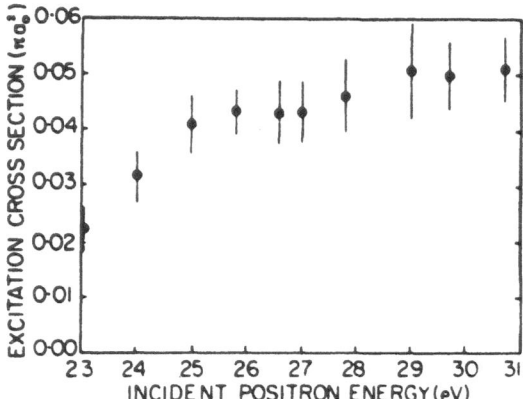

Fig. 8.   Partial e[+]-He excitation cross section. (From Coleman
          and Hutton, Ref. 21)

positron-helium excitation cross sections measured by Coleman and
Hutton are shown in Fig. 8. At incident positron energies above 30
eV a secondary peak associated with ionization overlaps the
excitation peak making it difficult to assign excitation cross
sections. More recently Coleman et al.[22] have extended their TOF
work and have obtained lower bounds on "excitation plus ionization"
cross sections for He, Ne, and Ar, which correspond to scattering in
the forward direction.

    A summary of the low energy positron inelastic scattering
measurements for He and Ar is provided[23] in Fig. 9. The curves
shown in this figure are portions of the total cross section curves
(solid lines) and the extrapolated elastic cross section curves
(dashed lines) from Fig. 5. The increase in these curves due to
inelastic scattering should be reliable as a consequence of the
ability of the Detroit experiments[13,14] to discriminate 100% against
inelastically scattered positrons. The relative ortho-Ps formation
measurements of Charlton et al.[18] (open circles) have been added to
the extrapolated elastic curve and normalized to match the total
cross section curve at the respective excitation thresholds of He
and Ar. The partial excitation (asterisks) and partial "excitation
plus ionization" (pluses) cross section measurements of Coleman et
al.[22] have also been added to the extrapolated elastic cross section
curves. It is clear from Fig. 9 that if the extrapolations of the
elastic scattering cross sections above the Ps formation thresholds
are valid, then a large part of the inelastic scattering cross
section for He and Ar is unaccounted for by the present inelastic
scattering measurements.

    An investigation of inelastic scattering of intermediate energy
(20-500 eV) positrons by helium has been made by Griffith et al.[24]

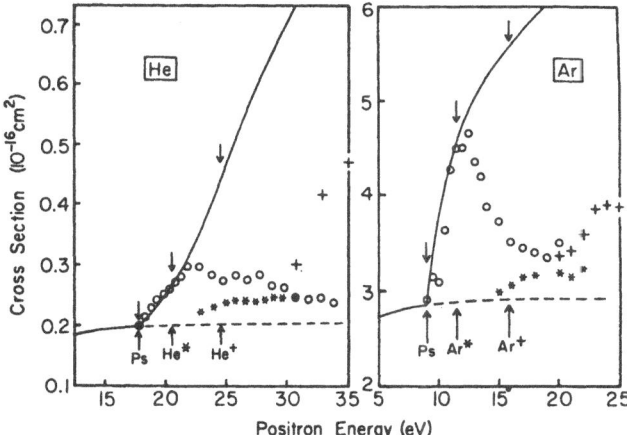

Fig. 9. Inelastic cross sections for e^+-He and e^+-Ar scattering. (From Stein and Kauppila, Ref. 23)

using the TOF approach. In general, they found a broad secondary peak in their TOF spectra associated with inelastic scattering. A significant feature of their measurements is that the energy loss corresponding to the maximum of their secondary peak increases with energy, which they feel implies that ionization is the dominant inelastic process between 100 and 500 eV. They also found that the fraction of timed scattered positrons to the total number of scattered positrons (timed and untimed) approached unity at higher energies which they interpret as implying that differential scattering is peaked in the forward direction and positronium formation must be insignificant at higher energies.

RESONANCE SEARCHES

Resonances have been predicted to exist for positron scattering by atomic hydrogen just below the n = 2 atomic excitation threshold[25] and associated with the first excited state of Ps formation.[26] There is also some theoretical evidence for the possible existence of a resonance for positron-helium scattering near 20.375 eV (above the Ps formation threshold and just below the first singlet excitation of He).[27] Up to the present time, there have not been any experimental observations of positron scattering resonances.

In our laboratory we have devoted some effort to searching for positron scattering resonances with our relatively narrow energy width (<0.1 eV) positron beam in a beam transmission experiment. In Fig. 10 are displayed measurements of the transmitted beam current versus the voltage applied to the positron source over 1.0 volt

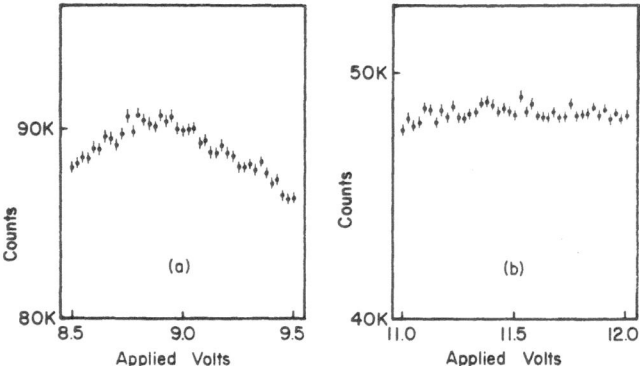

Fig. 10.   Resonance searches in e$^+$-Ar scattering. (From Stein et
           al., Ref. 28)

ranges centered near the Ps formation threshold (9.0 eV) and the
lowest atomic excitation threshold (11.5 eV) for Ar.[28] These
results were taken at 25-meV intervals with the primary beam
attenuated by about 50%. In this data there is no convincing
evidence of a resonance (which would be expected to manifest itself
as a relatively narrow structure in the transmitted beam current).
There is also no evidence of resonances in preliminary searches that
we have made with He and H$_2$. The abrupt change in the slope of the
transmitted current for Ar in Fig. 10(a) reflects the abrupt
increase in the total cross secton for Ar at the Ps formation
threshold (see Fig. 5), while the relatively constant slope observed
in Fig. 10(b) is consistent with the smooth shape of the total cross
section curve in the vicinity of 11.5 eV.

ADDED NOTE: Some recent measurements of 2^1S excitation and
ionization cross sections for positron-He scattering have been
reported by Sueoka,[29] who used a retarding potential-TOF method.

ACKNOWLEDGEMENTS

We wish to thank Prof. Wilhelm Raith for providing information
on some of his current research efforts. The Wayne State
positron-gas scattering research program is supported by the
National Science Foundation (Grant PHY80-07984).

REFERENCES

1. W. Raith, preceding paper presented at this Workshop.
2. P.G. Coleman and J.D. McNutt, Phys. Rev. Lett. 42:1130 (1979).

3. P.G. Coleman, J.D. McNutt, J.T. Hutton, L.M. Diana, and J.L. Fry, Rev. Sci. Instrum. 51:935 (1980).

4. D.M. Schrader, Phys. Rev. A 20:918 (1979).

5. R.P. McEachran, A.G. Ryman, and A.D. Stauffer, J. Phys. B 12:1031 (1979).

6. J.F. Williams, J. Phys. B 12:265 (1979).

7. T.S. Stein and W.E. Kauppila, private communication.

8. J.P. Bromberg, in: "The Physics of Electronic and Atomic Collisions," J.S. Risley and R. Geballe, editors, University of Washington Press, Seattle (1975) p. 98.

9. A.W. Yau, R.P. McEachran, and A.D. Stauffer, J. Phys. B 11:2907 (1978).

10. R.K. Nesbet, J. Phys. B 12:L243 (1979).

11. W. Raith, private communication.

12. W.E. Kauppila and T.S. Stein, Can. J. Phys. 60:471 (1982).

13. T.S. Stein, W.E. Kauppila, V. Pol, J.H. Smart, and G. Jesion, Phys. Rev. A 17:1600 (1978).

14. W.E. Kauppila, T.S. Stein, and G. Jesion, Phys. Rev. Lett. 36:580 (1976).

15. M.S. Dababneh, W.E. Kauppila, J.P. Downing, F. Laperriere, V. Pol, J.H. Smart, and T.S. Stein, Phys. Rev. A 22:1872 (1980).

16. W.E. Kauppila, T.S. Stein, J.H. Smart, M.S. Dababneh, Y.K. Ho, J.P. Downing, and V. Pol, Phys. Rev. A 24:725 (1981).

17. M.S. Dababneh, Y.-F. Hsieh, W.E. Kauppila, V. Pol, and T.S. Stein, Phys. Rev. A 26:1252 (1982).

18. M. Charlton, T.C. Griffith, G.R. Heyland, K.S. Lines, and G.L. Wright, J. Phys. B 13:L757 (1980).

19. D.R. Cook, P.G. Coleman, L.M. Diana, and S.C. Sharma, in: "Positron Annihilation," P.G. Coleman, S.C. Sharma, and L.M. Diana, editors, North-Holland, Amsterdam (1982) p. 87.

20. W.E. Kauppila and T.S. Stein, private communication.

21. P.G. Coleman and J.T. Hutton, Phys. Rev. Lett. 45:2017 (1980).

22. P.G. Coleman, J.T. Hutton, D.R. Cook, L.M. Diana, and S.C. Sharma, Proc. 12th Int. Conf. Phys. Electron. Atom. Collisions Abstracts p. 426 (1981).

23. T.S. Stein and W.E. Kauppila, Adv. Atom. Mol. Phys. 18:53 (1982).

24. T.C. Griffith, G.R. Heyland, K.S. Lines, and T.R. Twomey, J. Phys. B 12:L747 (1979).

25. G.D. Doolen, J. Nuttall, and C.J. Wherry, Phys. Rev. Lett. 40:313 (1978).

26. G.D. Doolen, Int. J. Quant. Chem. 14:523 (1978).

27. Y.K. Ho and P.A. Fraser, J. Phys. B 9:3213 (1976).

28. T.S. Stein, F. Laperriere, M.S. Dababneh, Y.-F. Hsieh, V. Pol, and W.E. Kauppila, Proc. 12th Int. Conf. Phys. Electron. Atom. Collisions Abstracts p. 424 (1981).

29. O. Sueoka, J. Phys. Soc. Jpn. 51:3757 (1982).

SURVEY OF RECENT THEORETICAL RESULTS ON POSITRON

SCATTERING IN GASES: LOW ENERGY

R.P. McEachran

Physics Department
York University
Toronto Canada

INTRODUCTION

The theoretical aspects of low energy positron-atom and positron-molecule scattering have been recently thoroughly reviewed by Drachman (1982) and by Schrader and Svetic (1982). This article will therefore deal primarily with the new theoretical approaches and calculations which have occurred in the past two years; as well, where it is known to the author, an outline of work still in progress will be mentioned.

ATOMIC HYDROGEN

In the pure elastic scattering region, $k^2 < \frac{1}{2}$ (where k^2 is the energy of the incident positron in Rydbergs), effectively 'exact' results exist for the various phase shifts eg. Bhatia et al (1971) for the S-wave and Bhatia et al (1974) for the P-wave.

The existence of these exact values continues to make positron-hydrogen atom collisions a testing ground for new theoretical approaches. An excellent measure of the reliability of any new set of phase shifts η, is given by the so-called "quality factor" (Drachman 1968)

$$Q = \left[\eta - \eta(\text{static})\right] / \left[\eta(\text{exact}) - \eta(\text{static})\right] \tag{1}$$

Recently, Pelikan and Klar (1983) have treated this system in hyperspherical coordinates and obtained S-wave phaseshifts in both an adiabatic and a post-adiabatic approximation. The quality factors, Q, for their results are shown in table 1; although they are close to 1 for some energies, they vary considerably and deteriorate close to

27

the positronium threshold. Elastic S-wave phase shifts have also been
determined (Darewych and McEachran 1981 and Horbatsch et al 1983a)
from the Hulthén-Kohn variational principle with a simple trial function of the form

$$\Psi_T = \frac{u(x)}{x}\ \psi_o(r) + \phi(x)\psi(r)\chi(s) \qquad (2)$$

Here x, r and s are the e^+ - proton, e^- - proton and e^+ - e^- distances
respectively. $\psi_o(r)$ is the hydrogen ground state wavefunction, while
u,ϕ,ψ and χ are functions for which coupled <u>differential equations</u> can
be obtained variationally. The quality factors, Q for these results
are approximately constant ($\sim$0.9) over the entire energy range and are
also given in table 1. Although there may be some numerical difficulties in solving differential equations of this type this method can, in
principle, be applied to larger systems. Ficocelli Varracchio (1983)
has developed a new approach which treats, in a unified fashion, both
subreactive and rearrangement processes for the e^+-H system. Here the
collision process is described by an optical potential which can be
written as the sum of three terms which represent respectively the
static potential, the non-adiabatic polarisation potential and the re-
arrangement channel. The quality factors Q, for a preliminary calculation including only the static and part of the polarisation potential
are also given in table 1.

Table 1. Drachman's quality factor, Q, equation (1).

k	(1)	(2)	(3)	(4)	(5)
0.1	0.669	0.858	0.170	0.834	0.882
0.2	0.608	0.929	0.041	0.869	0.912
0.3	0.511	0.963	0.039	0.891	0.929
0.4	0.376	0.852	0.077	0.906	0.932
0.5	0.197	0.645	0.183	0.917	0.934
0.6	-0.020	-	0.163	0.927	0.942
0.7	-0.270	-	0.391	0.935	0.950

(1) adiabatic approximation: Pelikan and Klar 1983
(2) post-adiabatic approximation: Pelikan and Klar 1983
(3) Ficocelli Varracchio (private communication)
(4) Darewych and McEachran 1981   (5) Horbatsch et al 1983a

For incident positron energies above $k^2=1/2$ and below $k^2 =3/4$
(corresponding to the n=2 excitation threshold) the only <u>inelastic</u>
process possible is positronium formation. The recent S-wave cross
sections for positronium formation of Humberston (1982), are based
upon a two-channel version of the Kohn variational method, using trial
functions containing up to 120 linear variational parameters, and are
the most reliable to date. They confirmed the earlier cross sections
of Stein and Sternlicht (1972) and showed that those of Chan and
Fraser (1973) were somewhat too small; these cross sections are given
in table 2. Nonetheless Humberston's S-wave results do not join
smoothly at $k^2 = 3/4$ onto the <u>total</u> inelastic S-wave cross section of
Winick and Reinhardt (1978). Humberston (private communication) has

now extended his work to the P-wave (but employing only 35 correlation terms) and has obtained positronium formation cross sections very similar to those of Chan and McEachran (1976); these values are also shown in table 2.   In the case of the P-wave both these results join smoothly onto the corresponding results of Winick and Reinhardt (1978). For higher partial waves, L>1, the most reliable results are still the Born approximation cross sections of Drachman et al (1976).

Table 2.   Positronium formation cross sections (πa^2_o)

k	S-wave			P-wave	
	(1)	(2)	(3)	(4)	(5)
0.71	0.0015	–	0.0041	0.0147	0.024
0.75	0.0029	0.0042	0.0044	0.357	0.33
0.80	0.0031	0.0048	0.0049	0.505	0.51
0.85	0.0032	–	0.0058	0.584	0.81

(1) Chan and Frader (1973)   (2) Stein and Sternlicht (1972)
(3) Humberston (1982)   (4) Chan and McEachran (1976)
(5) Humberston (private communication)

Choo et al (1978) showed that an infinite series of resonances exist below the n=2 threshold. The lowest of these resonances has been accurately determined by Doolen et al (1978) (their resonance parameters are $E_1 = 0.1003$eV and $\Gamma_1 = 1.84 \times 10^{-3}$eV). So far these resonances have not been reliably detected in a regular scattering calculation. However, Fraser (private communication) is currently solving a multi-channel scattering problem with an S-wave trial function of the form

$$\Psi_T(\bar{x},\bar{r}) = \psi_{1s}(r)F_{1s}(x) + \phi_{1s}(\rho)G_{1s}(\sigma) + \psi_{2s}(r)F_{2s}(x)$$
$$+ (\hat{r}\cdot\hat{x})\{\psi_{2p}(r)F_{2p}(x) + \bar{\bar{\psi}}_{2p}(r)F_{\overline{2p}}(x) + \bar{\phi}_{2p}(\rho)G_{\overline{2p}}(\sigma)\} \quad (3)$$

Here ψ_{1s}, ψ_{2s} and ψ_{2p} are the usual hydrogen wave functions, ϕ_{1s} is the ground state positronium wavefunction and $\bar{\bar{\psi}}_{2p}$ and $\bar{\phi}_{2p}$ are pseudostates chosen to give the correct long range dipole polarisabilities of hydrogen and positronium respectively. Although the absence of short range correlation terms in the trial function (3) will preclude highly accurate positronium formation cross sections its basic form should nonetheless enable it to detect the resonance structure below the n=2 threshold.   The generalization of this trial function to higher angular momentum is straight-forward.

Armour (1982) recently extended the proof that the e^+H system has no bound state to take into account the finite mass of the nucleus and has thereby shown (Armour 1983) that the systems $\mu^+e^-e^+$ and $p\mu^-e^+$ have no bound states.

NOBLE GASES

     The long-standing controversy concerning an apparent discrep-
ancy between theory and experiment with regard to the magnitude of
the elastic cross section near the Ramsauer minimum ($\sim$2eV) in He
appears to have been satisfactorily resolved once proper account has
been taken of the various angular discriminations in different
experiments (e.g. Stein and Kauppila 1982 and references cited
therein). Furthermore there is generally good agreement between
theory and experiment for the elastic scattering of positrons from
the other noble gases Ne, Ar, Kr and Xe. It will probably require
more accurate experiments based upon the introduction of intense
positron beams (e.g. Canter and Mills 1982) before a more detailed
assessment of the agreement between theory and experiment can be
given.

     In positron molecule collisions polarisation effects are
usually taken into account by means of phenomenological dipole polarisation potentials with adjustable parameters. Recently Horbatsch
et al (1983b) have developed a quantum-statistical polarised-density
approach which, in principle, is applicable to both atomic and molecular collisions and have tested this technique on neon and argon. In
this method the energy expression E for the adiabatic Hamiltonian is
expressed in terms of the electronic charge density $\rho(\overline{r}) = \psi^*(\overline{r})\psi(\overline{r})$
rather than in terms of the electronic wave function $\psi(\overline{r})$. The polarisation potential can then be written as

$$V_{pol}(x) = E[\rho(\overline{r}),\ x] - E[\rho_0(\overline{r}),x)] \tag{4}$$

where ρ_0 is the <u>unpolarised</u> charge density and x is the fixed position
of the positron. Here, instead of the Hartree-Fock expression for
the total electronic energy, one uses in (4) the extended Thomas-Fermi
energy functional (e.g. Gross and Dreizler, 1979). The charge density
$\rho(r)$ is then expanded in multipoles according to

$$\rho(\overline{r}) = \sum_{\ell=0}^{\infty} F_\ell(r;x)\ P_\ell(\cos\theta) \tag{5}$$

Horbatsch et al determine $F_\ell(r;x)$ in an approximate way: the functions $F_\ell(r;x)$ contain adjustable parameters which are determined
variationally by minimising the adiabatic energy $E[\rho(\overline{r}),\ x]$ for
each value of x. In table 3 we compare total elastic cross sections
for e^+ - Ne and e^+ - Ar collisions with a sample selection of other
theoretical and experimental results (the percentages given in
brackets are correctional estimates due to angular discrimination,
Stein and Kauppila (1982)).

Excitation

     Coleman et al (1982) reported excitation and ionization measure-

Table 3. Total elastic cross sections for e⁺-Ne and e⁺-Ar.(πa_0^2)

	Neon				Argon		
E(eV)	(1)	(2)	(3)		(1)	(2)	(3)
0.50	0.371	0.311	0.315(+10%)		12.65	2.96	7.18(+ 8%)
0.75	0.197	0.168	0.120(+15%)		8.37	2.33	4.27(+12%)
1.25	0.170	0.207	0.208(+ 9%)		5.23	2.30	3.22(+16%)
2.00	0.278	0.354	0.350(+ 4%)		3.89	2.69	2.73(+20%)
3.50	0.476	0.597	0.593(+2.5%)		3.60	3.13	2.98(+20%)
6.50	0.712	0.856	0.902(+ 2%)		3.21	3.42	3.19(+19%)

(1) Polarised orbital: McEachran et al (1978, 1979)
(2) Extended Thomas-Fermi: Horbatsch et al (1983b)
(3) Experiment: Stein et al (1978)
(4) Experiment: Kauppila et al (1976)

The agreement for Ne is very good whereas for Ar there is a discrepancy at low energy.

ments for positrons incident upon helium, neon and argon. Their results for the excitation of the 2^1S state of He are shown in table 4. Recently Parcell et al (1983) performed a theoretical calculation of this cross section using a distorted wave approximation (e.g. Madison 1979). The T matrix element for this interaction can be written exactly as $\langle \psi_{b1}^- | V | \psi_a^+ \rangle$. However, it can be approximated by $\langle \psi_{b1}^- | V | \psi_{a1}^+ \rangle$ where ψ_{a1}^+ is a solution of the equation

$$(H_o - V_1^a - E) \psi_{a1}^+ = 0 \qquad\qquad (6)$$

with outgoing boundary conditions describing the system in channel a. (ψ_a^+ is the solution of equation (6) with V_1^a replaced by V.) ψ_{b1}^- is a solution of the equation

$$(H_o + V_1^b - E) \psi_{b1}^- = 0 \qquad\qquad (7)$$

with incoming boundary conditions describing the system in channel b and $V = V_1^a + V_2^a = V_1^b + V_2^b$ is the interaction potential. Here the potentials V_1^a and V_1^b are functions of the positron position only and are chosen to best describe elastic scattering in the respective channels. In this calculation these potentials were chosen as the sum of a static and a polarisation potential. The polarisation potential for the gound state was taken from the polarised orbital calculation of McEachran et al (1977). Several polarisation potentials were tried in the excited state channel. In table 4 are presented the experimental results of Coleman et al (1982) together with two theoretical calculations, one with a scaled Dalgarno-Lynn polarisation potential

(1957) in the excited channel and the other with a model potential
as advocated by Schrader (1979). Both these potentials were constructed to yield asymptotically the correct dipole polarisability
($\sim$802 au) (Chung and Hurst 1966) of the 2^1S state of helium. The
agreement between theory and experiment is reasonable above 25eV
but deteriorates as one approaches the 2^1S threshold at 20.6eV.

Table 4. $1^1S - 2^1S$ excitation cross section in He ($\pi a_0{}^2$)

E_{inc}(eV)	(1)	(2)	(3)
21	–	0.0420	0.0451
22	–	0.0460	0.0545
23	0.022 ± 0.004	0.0442	0.0543
24	0.031 ± 0.005	0.0436	0.0537
25	0.040 ± 0.005	0.0437	0.0529
26	0.045 ± 0.004	0.0434	0.0520
27	0.042 ± 0.006	0.0427	0.0512
28	0.049 ± 0.005	0.0418	0.0503
29	0.050 ± 0.005	0.0412	0.0496

(1) Experiment: Coleman et al 1982
(2) Dalgarno-Lynn potential: Parcell et al 1983
(3) Schrader potential: Parcell et al 1983

MOLECULES

Most theoretical positron-molecule cross section calculations
can be characterized by the phenomenological form chosen for the
polarisation potential. Darewych (1982) employed the following form
of polarisation potential for $e^+ - N_2$ collisions, namely

$$V_{pol}(r) = - \frac{1}{2r^4} \{1-\exp[-(r/r_c)^6]\}\{\alpha_0 + \alpha_2 P_2(\cos\theta)\} \tag{8}$$

Here the polarisabilities α_0 and α_2 were chosen to be 12.0 and 4.2 au.
respectively (Burke and Chandra, 1972). It was found that satisfactory agreement with experiment (Hoffman et al (1982) and Charlton et al
1981) for the total cross section could not be achieved with a single
fixed value of the cut-off parameter r_c; however if r_c was allowed to
be energy dependent then the experimental cross section could be
reproduced as is shown in figure 1. Also shown in figure 1 are the
previous calculations of Darewych and Baille (1974) and Gillespie and
Thompson (1975) which can be approximately reproduced by fixing r_c to
be 1.59 and 2.2 respectively. Clearly, the Darewych and Baille
potential is too attractive at low energies and the Gillespie and
Thompson potential is evidently too weak at all energies. Darewych
(1982) also calculated $Z_{eff} = 17.5$ at thermal energies; this value is

somewhat below the corresponding experimental measurements of 40.0 $\pm$ 0.9 (Sharma and McNutt 1978) and 29.7 $\pm$ 0.2 (Coleman and Griffith 1973) as is characteristic of calculations which do not take account of correlation effects.

Recently, Horbatsch and Darewych (1983) have used this same form of phenomenological polarisation potential as given by equation (8) as well as the form

$$V_{pol}(r) = - \frac{1}{2r_>^4}\{\alpha_o + \alpha_2 P_2 (\cos\theta)\} \quad r_> = \max(r,r_c) \qquad (9)$$

(Takayangi and Geltman 1965; Schrader 1979) to calculate the total cross section for $e^+ - CO_2$ collisions. Here the polarisabilities α_o and α_2 were chosen to be 17.90 and 9.19 au respectively (Morrision et al 1977). They found that, with either form of potential, (8) or (9), if they fixed the value of r_c so as to reproduce experiment (Hoffman et al 1982, Griffith 1982) at 5eV then although they could achieve agreement with the London experiment, they were unable to reproduce the very low energy cross sections of the Wayne State group. On the other hand if r_c was choosen to be energy dependent then satisfactory agreement with this latter experiment could be achieved. These results are shown in figure 2.

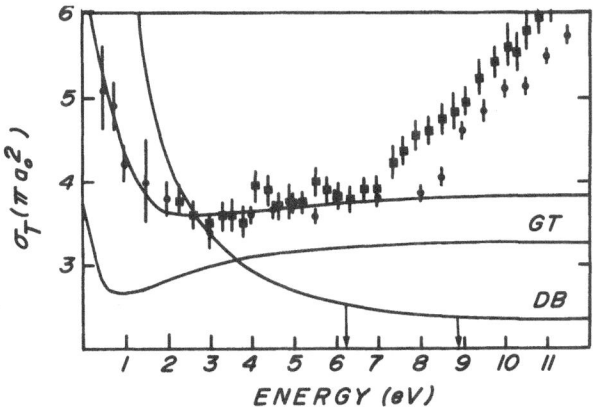

Fig. 1. $e^+ - N_2$ cross sections. ●, total cross section measurements of Hoffman et al (1982). ■, total cross section measurements of Charlton et al (1981). D, elastic cross section of Darewych (1982). GT, elastic cross section of Gillespie and Thompson (1975). DB, elastic cross section of Darewych and Baille (1974).

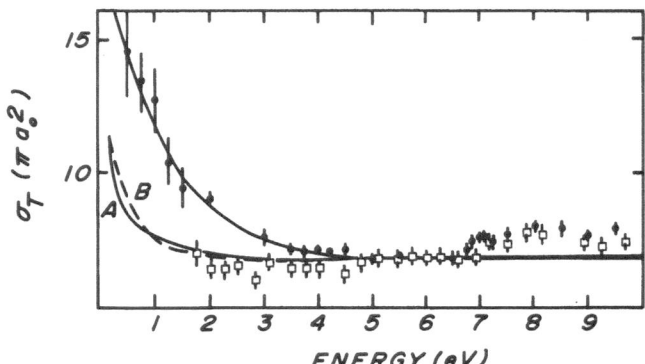

Fig. 2 $e^+ - CO_2$ cross sections. ●, total cross section measurements
of Hoffman et al (1982). ▢, total cross section measurements
of Griffith (1982). A,B elastic cross section for potentials
given by equations (8) and (9) respectively. Unlabelled
curve, elastic cross section for energy dependent cut-off
parameter.

Jain and Thompson (1983) have reported the first theoretical
calculations for $e^+ - CH_4$ and $e^+ - NH_3$ collisions. They tried three
forms for the polarisation potential, namely a parameter free potential calculated using the Pople-Schofield method (see Jain and
Thompson 1982) as well as the functional forms

$$V_{pol}(r) = -\frac{\alpha_o}{2r^4}\left\{1 - \exp\left[-(r/r_c)^6\right]\right\}, \quad r_c = 2.066 \text{ au} \qquad (10)$$

$$V_{pol}(r) = -\frac{\alpha_o}{2r^4}\left[1 - \exp(-r/r_c)\right]^6, \quad r_c = 0.88 \text{ au} \qquad (11)$$

Their total cross sections for CH_4 obtained with these three potentials are shown in figure 3 together with the recent experimental
results of Charlton et al (1983). Once again, it is seen that polarisation potentials with fixed cut-off parameters r_c do not yield reliable results; on the other hand the cross section determined with
their parameter free polarisation potential is in quite satisfactory
agreement with experiment. Jain and Thompson (1983) obtain a value of
$Z_{eff} = 99.5$ at thermal energies which is somewhat lower than the experimental values of 139.6 ± 1.0 (Smith and Paul 1970) and 153.7 ± 0.9 (McNutt
et al 1975). For NH_3 their corresponding value of Z_{eff} is 241 which
is considerably lower than the experimental result of 1300 (Heyland
et al 1982) and they postulate the formulation of a cluster of NH_3
molecules around the positron to account for this large value.

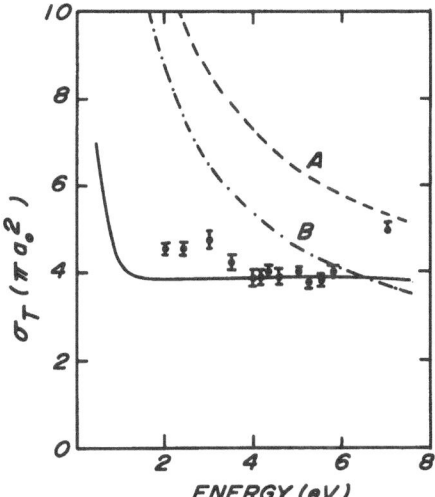

Fig. 3. e^+ - CH cross sections. ●, total cross section of Charlton et al (1983). A, B, elastic cross section for potentials given by equations (10) and (11) respectively. Unlabelled curve, elastic cross section for parameter free potential.

Currently Armour and Lavender (private communication) are performing an elaborate calculation using confocal elliptical coordinates to determine the cross section for the elastic scattering of low energy positrons by the hydrogen molecule. This calculation is based upon the Kohn variational method and their trial wavefunction is a generalisation (to include short range closed channel functions) of the wavefunction originally proposed by Massey and Ridley (1956). Preliminary calculations have been carried out using 22 closed channel functions and are in qualitative agreement with experiment.

REFERENCES

Armour EAG 1982 Phys. Rev. Lett. 48 1578-81
Armour EAG 1983 J. Phys. B: At. Mol. Phys. 16 1295-302
Bhatia AK, Temkin A, Drachman R and Eiserike H 1971 Phys. Rev. A3
    1328-35
Bhatia AK, Temkin A and Eiserike H 1974 Phys. Rev. A9 219-22

Burke PG and Chandra N 1972 J. Phys. B: At. Mol. Phys. 5 1696-711
Canter KF and Miles Jr, AP 1982 Can. J. Phys. 60 551-7
Chan YF and Fraser PA 1973 J. Phys. B: At. Mol. Phys. 6 2504-15
Chan YF and McEachran RP 1976 J. Phys. B: At. Mol. Phys. 9 2869-75
Charlton M, Griffith TC, Heyland GR and Wright GL 1981 Eur. Conf.
     on Atomic Physics, Heidelberg Abstracts ed J. Kowalski,
     G zu Putlitz and HG Weber (Heidelberg: European Physical
     Society) p 696
Charlton M, Griffith TC, Heyland GR and Wright GL 1983 J. Phys. B:
     At. Mol. Phys. 16 323-41
Choo LT, Crocker MC and Nuttall J 1978 J. Phys. B: At. Mol. Phys.
     11 1313-22
Chung KT and Hurst RP 1966 Phys. Rev. 152 35-41
Coleman PG and Griffith TC 1973 J. Phys. B: At. Mol. Phys. 6 2155-61
Coleman PG and Hutton JT, Cook DR and Chandler CA 1982 Can. J. Phys.
     60 584-90
Dalgarno A and Lynn N 1957 Proc. Phys. Soc. (London) A70 223-5
Darewych JW 1982 J. Phys. B: At. Mol. Phys. 15 L415-9
Darewych JW and Baille P 1974 J. Phys. B: At. Mol. Phys. 7 L1-4
Darewych JW and McEachran RP 1981 J. Phys. B: At. Mol. Phys. 14
     4415-24
Doolen GD, Nuttall J and Wherry CJ 1978 Phys. Rev. Lett. 40 313-5
Drachman RJ 1968 Phys. Rev. 173 190-202
Drachman RJ 1982 Can. J. Phys. 60 494-502
Drachman RJ Omidvar K and McGuire JH 1976 Phys. Rev. A 14 100-3
Ficocelli Varracchio E 1983 Ann. Phys. 145 131-61
Gillespie ES and Thompson DG 1975 J. Phys. B: At. Mol. Phys. 8
     2858-68
Griffith TC 1982 (private communication to JW Darewych)
Gross EKU and Dreizler PM 1979 Phys. Rev. A20 1798-807
Heyland GR, Charlton M, Griffith TC and Wright GL 1982 Can. J. Phys.
     60 503-16
Hoffman KR, Dababneh MS, Hsieh Y-F, Kauppila WE, Pol V, Smart JH and
     Stein TS 1982 Phys. Rev. A25 1393-403
Horbatsch M and Darewych JW 1983 J. Phys. B: At. Mol. Phys.
     (accepted)
Horbatsch M, Darewych JW and McEachran RP 1983a J. Chem. Phys.
     (accepted)
_____ 1983b J. Phys. B: At. Mol. Phys. (accepted)
Humberston JW 1982 Can. J. Phys. 60 591-6
Jain A and Thompson DG 1982 J. Phys. B: At. Mol. Phys. 15 L631-7
_____ 1983 J. Phys. B: At. Mol. Phys. 16 1113-23
Kauppila WE, Stein TS and Jesion G 1976 Phys. Rev. Lett 36 580-4
Madison DH 1979 J. Phys. B: At. Mol. Phys. 12 3399-414
Massey HSW and Ridley RO 1956 Proc. Phys. Soc. A69 659-67
McEachran RP, Morgan DL, Ryman AG and Stauffer AD 1977 J. Phys. B:
     At. Mol. Phys. 10 663-77
McEachran RP, Ryman AG and Stauffer AD 1978 J. Phys. B: At. Mol.
     Phys. 11 551-61
_____ 1979 J. Phys. B: At. Mol. Phys. 12 1031-41

McNutt JD, Summerour DA, Ray AD and Huang PH 1975 J. Chem. Phys. 62
      1777-89
Morrison MA, Lane NF and Collins LA 1977 Phys. Rev. A15 2186-201
Parcell LA, McEachran RP and Stauffer AD 1983 J. Phys. B: At. Mol.
      Phys. (accepted)
Pelikan E and Klar H 1983 Z. fur Phys. A, Atoms and Nuclei 310 153-8
Schrader DM 1979 Phys. Rev. 20 918-32
Schrader DMand Svetic RE 1982 Can. J. Phys. 60 517-42
Sharma SC and McNutt JD 1978 Phys. Rev. A18 1426-34
Smith PM and Paul DAL 1970 Can. J. Phys. 48 2984-90
Stein J and Sternlicht R 1972 Phys. Rev. A 6 2165-9
Stein TS and Kauppila WE 1982 Advances in Atomic and Molecular
        Physics Vol 15 (New York: Academic) pp 53-96
Stein TS, Kauppila WE, Pol V, Smart JH and Jesion G 1978 Phys. Rev.
      A 17 1600-8
Takayanagi K and Geltman S 1965 Phys. Rev. 138 1003-10
Winick JR and Reinhardt WP 1978 Phys. Rev. A18 925-34

SURVEY OF RECENT THEORETICAL RESULTS ON POSITRON

SCATTERING IN GASES : INTERMEDIATE AND HIGH ENERGIES

C.J. Joachain

Physique Théorique
Université Libre de Bruxelles
Belgium

1. INTRODUCTION

Positron-gas collision experiments are of great interest not only because they yield direct information on the interactions of positrons with atoms and molecules, but also because they can help us to understand better electron-atom (molecule) scattering. A review of the first decade of activity in the field of e^+-gas experiments, which started in 1972, has been given by Stein and Kauppila (1982). In addition to total cross section measurements, a new generation of experiments aiming at the determination of integrated cross sections for elastic scattering, excitation, ionization and positronium formation as well as some differential cross sections have either been performed or are being planned. In this article I shall give a survey of recent theoretical work dealing with these processes, for incident positron energies well above the ionization energy of the target, but non-relativistic. Atomic units (a.u.) will be used.

2. ELASTIC SCATTERING AND TOTAL CROSS SECTIONS

Let us begin by considering the simplest collision process, namely elastic scattering, and assume that the wave number k of the incident positron (which is equal in a.u. to its velocity v) is large enough so that perturbation theory can be applied. In the limit $k \gg 1$ the first Born amplitude $\bar{f}_{B1}$, which yields identical cross sections for electrons and positrons, governs the scattering at all angles. If k is decreased, while remaining somewhat larger than unity, higher order methods are required to obtain accurate results and to explore the differences between e^+ and e^- scattering.

The Glauber approximation (Glauber, 1959) gives identical e^+ and e^- cross sections and suffers from severe deficiencies (Byron and Joachain, 1973, 1977a, Joachain and Quigg, 1974) so that it cannot be used directly in this context. However, one can use the Eikonal-Born series method (Byron and Joachain, 1973, 1975, 1977a) and construct the amplitudes

$$f_{EBS} = \overline{f}_{B1} + \overline{f}_{B2} + \overline{f}_{G3} \tag{1}$$

or

$$f'_{EBS} = f_G - \overline{f}_{G2} + \overline{f}_{B2}$$
$$= \overline{f}_{B1} + \overline{f}_{B2} + \sum_{n=3}^{\infty} \overline{f}_{Gn} \tag{2}$$

where $\overline{f}_{Bn}$ is the term of order n of the Born series, f_G is the Glauber amplitude and $\overline{f}_{Gn}$ is the term of order n of the Glauber series, obtained by expanding f_G in powers of the interaction potential acting between the projectile and the target. Both EBS amplitudes (1) and (2) are consistent through order k^{-2} and eliminate the serious deficiencies of f_G, but they still suffer from a lack of all-order unitarity.

In order to improve on this point, Byron, Joachain and Potvliege (1981, 1982) have recently proposed to unitarize the EBS method in the following way. First, the eikonal amplitude of Wallace (1973), which contains in a systematic way the corrections to the eikonal phase, is generalized to the multiparticle case. The resulting many-body Wallace amplitude f_W is still a zero-excitation-energy approximation and therefore it does not account for long-range polarization effects at small angles and it represents inadequately absorption effects in the same region. Fortunately, these difficulties can be eliminated by removing the term $\overline{f}_{W2}$ of the many-body Wallace approximation which is of second order in the projectile-target interaction, and replacing it by $\overline{f}_{B2}$. The direct amplitude thus obtained,

$$f_{UEBS} = f_W - \overline{f}_{W2} + \overline{f}_{B2} \tag{3}$$

is nearly unitary (to all orders) at all angles. It is called the Unitarized Eikonal-Born series amplitude. At small momentum transfers, where higher-order terms of perturbation theory are rather unimportant, one retrieves the EBS amplitude (1) by keeping terms through order k^{-2} in (3). On the other hand, at large momentum transfers, f_{UEBS} provides more accurate values than the EBS amplitude (2).

The Unitarized EBS method which has been described above can be applied to inelastic as well as elastic collisions. Another way of unitarizing the EBS method, which has been used for elastic scattering, is to transform it into an optical model approximation

(Byron and Joachain, 1974, 1977a,b, 1981, Joachain et al., 1977). In this approach the lowest order terms of perturbation theory, calculated by using the EBS method, are converted into an optical potential which is then treated in an unitary, partial-wave manner. This retains all the advantages of the EBS theory at small and intermediate angles, but includes approximations to higher order terms of perturbation theory which are important at large angles, where the effects of the singular Coulomb potential are most important. Third order optical model (OM) calculations of this kind have been performed recently for $e^{\pm}$-H elastic scattering by Byron and Joachain (1981), while second order optical model calculations have been carried out for helium and neon (Byron and Joachain, 1974, 1977b) and argon (Joachain et al., 1977). Closely related to this approach is the second order potential (SOP) method developed by Bransden et al. (see for example Winters et al., 1974; Scott and Bransden, 1981) which has been applied by Mukherjee and Sural (1982) to e^{+}-H and e^{+}-He elastic scattering.

The angular distributions for elastic scattering of fast positrons by atomic hydrogen, helium, neon and argon have been discussed by Byron and Joachain (1977a,b,c). The main difference with electron scattering is that at small angles the strong forward peak characteristic of electron scattering is significantly reduced for positron scattering. This is primarily due to the fact that in the case of electron scattering the first Born term and the real part of the second Born term add constructively, while in the case of positron scattering they add destructively.

Turning now to total (integrated) elastic cross sections, we see from Table 1 that the four theoretical methods discussed above

Table 1. Total (integrated) elastic cross sections (in units of πa_o^2) for positron scattering by atomic hydrogen. FBA refers to the first Born approximation and EBS, UEBS, OM and SOP to the four methods discussed in the text. The numbers in parentheses indicate powers of ten.

E (eV)	FBA	EBS (1)	UEBS (2)	OM (3)	SOP (4)
100	2.99(-1)	2.43(-1)	2.28(-1)	2.23(-1)	2.70(-1)
200	1.54(-1)	1.34(-1)	1.31(-1)	1.31(-1)	1.38(-1)
300	1.04(-1)	9.39(-2)	9.28(-2)	9.29(-2)	–
400	7.83(-2)	7.23(-2)	7.18(-2)	7.19(-2)	–

(1) Byron and Joachain, 1977c
(2) Byron, Joachain and Potvliege, 1982
(3) Byron and Joachain, 1981
(4) Mukherjee and Sural, 1982

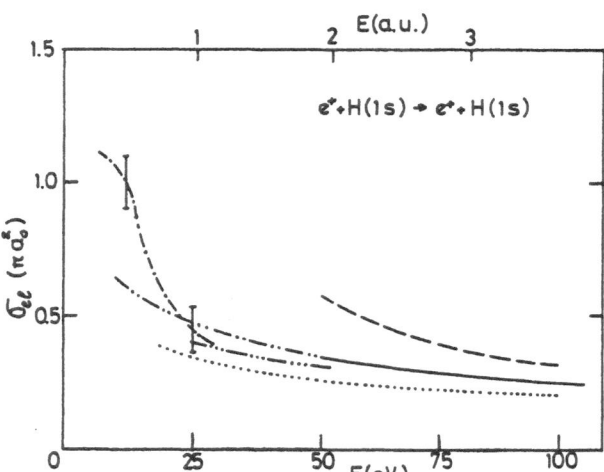

Fig. 1. Total (integrated) elastic cross section for the scattering
of positrons by atomic hydrogen. ——, third order optical
model (Byron and Joachain, 1981); ----, first Born approximation ; —·—·, moment T-matrix results of Winick and
Reinhardt (1978), with indication of possible errors;
—··—, three state close-coupling results and —···—,
twelve state close-coupling results of Morgan (1982);
.... results of Das and Biswas (1981).

are in excellent agreement for e^+-H scattering at energies E $\geqslant$ 200
eV, giving a cross section at 400 eV which is still 8% lower than
the first Born result. At 100 eV, however, they split, with the EBS
and especially the SOP results being on the higher side.

The situation at lower energies is illustrated in Fig. 1,
where the third-order optical model total elastic e^+-H cross section
along with the first Born result is compared with the values
obtained by Winick and Reinhardt (1978) using the moment T-matrix
approach, the three-state and twelve-state close-coupling results
of Morgan (1982) and the calculations of Das and Biswas (1981).

The total (integrated) cross sections for fast positron-
helium elastic scattering obtained from the EBS, the second order
optical model (OM) and the SOP method are given in Table 2 in the
energy range 100-500 eV, together with the first Born values and

Table 2. Total (integrated) elastic cross sections (in units of πa_o^2)
          for positron scattering by helium.

E (eV)	FBA	EBS (1)	OM (2)	SOP (3)	3CC (4)
100	4.11(−1)	3.66(−1)	1.72(−1)	2.21(−1)	2.42(−1)
200	2.20(−1)	1.61(−1)	1.31(−1)	1.58(−1)	1.62(−1)
300	1.52(−1)	1.15(−1)	1.04(−1)	1.10(−1)	−
400	1.15(−1)	9.17(−2)	8.63(−2)	−	−
500	9.29(−2)	7.64(−2)	7.32(−2)	−	−

(1) Byron and Joachain, 1977c
(2) Byron and Joachain, 1977b
(3) Mukherjee and Sural, 1982
(4) Willis, Hata, McDowell, Joachain and Byron, 1981.

the results of the three-state close-coupling calculations of
Willis et al. (1981) at 100 eV and 200 eV. The agreement between
the EBS, OM and SOP results at the higher energies is good, but as
the energy decreases significant differences appear, indicating in
particular the progressive breakdown of the EBS perturbation
treatment.

Let us now consider total (complete) positron cross sections
σ_{tot}^+. In Table 3 are given the values of σ_{tot}^+ for e^+-H collisions,
in the energy range 100-400 eV, as obtained by using the optical
theorem,

$$\sigma_{tot}^+ = \frac{4\pi}{k} \ \text{Im} \ f_{e\ell}^+(\theta=o), \qquad (4)$$

the imaginary part of the forward elastic amplitude $\text{Im} \ f_{e\ell}^+(\theta=o)$,
being given respectively by the EBS, the UEBS and the third order

Table 3. Total cross sections (in units of πa_o^2) for positron-
          atomic hydrogen scattering.

E (eV)	EBS (1)	UEBS (2)	OM (3)
100	2.38	2.23	2.15
200	1.40	1.34	1.32
300	1.01	9.78(−1)	9.64(−1)
400	7.96(−1)	7.77(−1)	7.70(−1)

(1) Byron and Joachain, 1977c
(2) Byron, Joachain and Potvliege, 1982
(3) Byron and Joachain, 1981

optical model theory. In the original EBS method the total cross section σ_{tot}^{+} and the corresponding electron quantity σ_{tot}^{-} are equal at a fixed energy since the imaginary part of the elastic amplitude is just $\mathrm{Im}\ \overline{f}_{B2}$, the imaginary part of the second Born term. In fact, the calculation of the difference $\sigma_{tot}^{-} - \sigma_{tot}^{+}$ is a very delicate test of theoretical methods, since this quantity depends on third and higher odd order contributions to the imaginary part of the direct elastic forward amplitude , $\mathrm{Im}\ f_{e\ell}(\theta=o)$, together with contributions to the imaginary part of the exchange elastic forward amplitude, $\mathrm{Im}\ g_{e\ell}(\theta=o)$. It is interesting to note that in the UEBS method the two quantities $\mathrm{Im}\ f_{e\ell}^{-}(\theta=o)$ and $\mathrm{Im}\ f_{e\ell}^{+}(\theta=o)$ are equal for $e^{\pm}$-H elastic scattering. As a result, the difference between σ_{tot}^{-} and σ_{tot}^{+} is due only to the exchange amplitude and remains quite small down to relatively low incident energies, the quantity $(\sigma_{tot}^{-} - \sigma_{tot}^{+})/\sigma_{tot}^{-}$ being equal to 0.025 at 100 eV.

A very detailed comparison between measurements of total cross sections for intermediate-energy positrons colliding with helium, neon and argon and theoretical calculations has been made by Kauppila et al. (1981) and Stein and Kauppila (1982), who have also compared the positron total cross sections with the corresponding electron ones. These comparisons will therefore not be repeated in this article. It is worth stressing, however, that the experiments of Kauppila et al. (1981), in which σ_{tot}^{-} and σ_{tot}^{+} are measured in the same apparatus, reveal that for $e^{\pm}$-He collisions the σ_{tot}^{-} and σ_{tot}^{+} curves merge at a relatively low energy (200 eV), a situation similar to that predicted for $e^{\pm}$-H collisions by the UEBS method, as described above. On the other hand, for neon and argon the positron and electron total cross section curves are slowly approaching each other at the higher energies.

Another important aspect of total $e^{\pm}$-atom cross sections studies, which has attracted considerable interest in recent years, is the question of the validity of the zero-energy sum rule based on the forward dispersion relations proposed by Gerjuoy and Krall (1960). For scattering by inert gases, this sum rule has the form (Bransden and McDowell, 1969)

$$- A - \overline{f}_{B1}(\theta=o) + \overline{g}_{B1}(\theta=o) = \frac{1}{2\pi} \int_{0}^{\infty} \sigma_{tot}(k)\ dk \qquad (5)$$

where A is the scattering length, $\overline{f}_{B1}(\theta=o)$ and $\overline{g}_{B1}(\theta=o)$ are respectively the first Born elastic scattering amplitudes in the forward direction for direct and exchange scattering, and σ_{tot} is the total cross section in units of πa_{0}^{2}. Tests of the sum rule (5) have been thoroughly discussed by Kauppila et al. (1981) and Stein and Kauppila (1982), who conclude, in agreement with earlier studies (Byron, de Heer and Joachain, 1975, Bransden and Hutt, 1975, de Heer et al., 1976, Hutt et al., 1976) that the sum rule (5) is valid for positron scattering but fails for electron scattering.

The failure of forward dispersion relations for electron-atom collisions has been studied by Byron, de Heer and Joachain (1975), de
Heer et al. (1976), Hutt et al. (1976), Blum and Burke (1977), Tip
(1977 a, b) and Byron and Joachain (1978) and is understood to arise
from the analytic structure of the forward elastic exchange amplitude, $g_{e\ell}$, as a function of the energy of the projectile electron.

Finally, we remark that Srivastava and Pathak (1981) have used
the EBS method together with the independent-atom model to obtain
total cross sections for e^{+}-H_2 collisions at intermediate and high
energies. Their results are in very good agreement with the experimental data of Charlton et al. (1980).

3. EXCITATION AND IONIZATION

Let us now turn to excitation collisions. In this case the
situation is more complicated than for elastic scattering, since
the first Born amplitude $\bar{f}_{B1}$ falls off very rapidly with increasing
momentum transfer, so that large angle scattering is dominated by
the second Born term $\bar{f}_{B2}$ at high energies. Byron, Joachain and
Potvliege (1981) have applied their UEBS method to the excitation
of the 2s state of atomic hydrogen by positron impact. The differential cross section they have obtained at 100 eV is shown in Fig. 2,
where it is compared with the corresponding UEBS results for electron
impact, and the first Born values. In contrast with the elastic
scattering case, the positron differential cross section is seen to
be larger than the electron one at small angles, and this causes the
total (integrated) 1s-2s cross section to be larger for positrons
than for electrons. A similar result has been obtained at lower
energies by Morgan (1982), who found that the close-coupling results
for the 1s-2s positron impact integrated cross section lie significantly higher than the first Born results. Moreover, the inclusion
of pseudostates increases the positron 1s-2s cross section, in
contrast to the electron case, where the pseudostates reduce the
1s-2s cross section significantly. Morgan also found that the
inclusion of pseudostates enhances the positron 1s-2p integrated
cross section but that the effect is much smaller in that case.

Calculations of differential and integrated cross sections for
excitation of the $1^{1}S \rightarrow 2^{1}S$ and $1^{1}S \rightarrow 2^{1}P$ transitions in helium by
positrons having energies between 60 and 400 eV have been performed
by Willis et al. (1981), using a three-state close-coupling (3CC)
model and also the EBS method and a distorted wave Born approximation.
Their 3CC results in the energy range 60-200 eV are given in Table 4,
where they are compared with the corresponding electron cross sections. We see from Table 4 that the positron cross sections are
higher than the electron ones for the $1^{1}S \rightarrow 2^{1}S$ transition, but lower
for the $1^{1}S \rightarrow 2^{1}P$ transition, the sum $\sigma(1^{1}S \rightarrow 2^{1}S) + \sigma(1^{1}S \rightarrow 2^{1}P)$ being
nearly independent of the sign of the projectile charge for incident

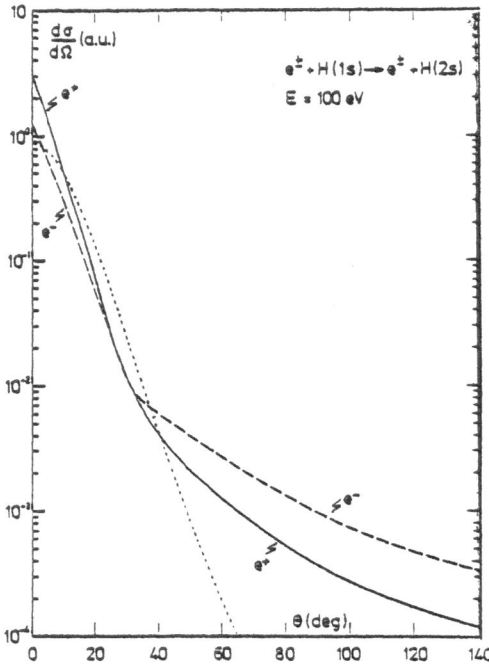

Fig. 2. The excitation of atomic hydrogen to the 2s state by posi-
        tron and electron impact at an incident energy of 100 eV.
        ———, UEBS calculation (Byron, Joachain and Potvliege, 1981)
        for positron impact, ---, UEBS results for electron impact,
        .... first Born approximation.

Table 4. Total (integrated) cross sections (in units of πa_0^2) for
         positron and electron excitation of the $1^1S \rightarrow 2^1S$ and
         $1^1S \rightarrow 2^1P$ transitions in helium, as obtained from the
         three-state close-coupling calculations of Willis et al.
         (1981).

E (eV)	$1^1S \rightarrow 2^1S$ transition e^+	e^-	$1^1S \rightarrow 2^1P$ transition e^+	e^-
60	3.49(-2)	2.84(-2)	1.28(-1)	1.80(-1)
80	3.09(-2)	2.28(-2)	1.39(-1)	1.62(-1)
100	2.67(-2)	1.91(-2)	1.39(-1)	1.48(-1)
150	1.88(-2)	1.39(-2)	1.23(-1)	1.22(-1)
200	1.43(-2)	1.09(-2)	1.09(-1)	1.04(-1)

energies E ⩾ 100 eV. Calculations of differential and integrated
cross sections for the excitation of the $1^1S \rightarrow 2^1S$ transition in
helium by low and intermediate energy positrons, using a distorted

wave approach, have also been performed by Parcell, McEachran and
Stauffer (1983).

Willis et al. (1981) have analyzed the coincidence parameters
(Eminyan et al., 1974) λ and $|\chi|$ for positron and electron impact
excitation of the 2^1P level at 80 eV. The positron results behave
very differently from the electron ones, which in turn implies quite
different behaviours of the linear and circular polarizations of the
2^1P $\rightarrow 1^1$S radiation in the e^+ and e^- cases. Experimental verifica-
tion, however, will have to await the availability of more powerful
positron sources.

Most of the detailed knowledge concerning electron impact
ionization of atoms has been gained during the recent years by ana-
lyzing $(e^-,2e^-)$ coincidence experiments; (e^+,e^+e^-) experiments performed with incident positrons would also yield very valuable information about ionization theory. As an example, the first and second
Born triple differential cross sections (TDCS) for $e^{\pm}$-H(1s) ionization, obtained by Byron, Joachain and Piraux (1980) are shown in
Fig. 3 for the case of a coplanar asymmetric, Ehrhardt-type

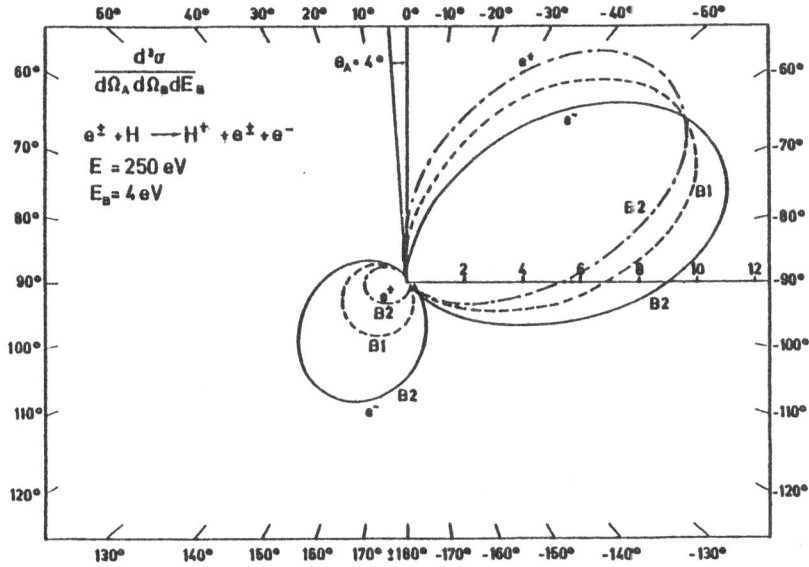

Fig. 3. The TDCS (in a.u.) for the ionization of atomic hydrogen by
electron and positron impact, for the case E = 250 eV,
θ_A = 4° and E_B = 4 eV. B1 (---) : first Born approximation;
B2 : second Born approximation for electrons (——) and
positrons (—.—.) calculated by using an average excitation
energy $\overline{w}$ = 0.5 a.u. All curves are normalized to the first
Born value at θ_B = − 60°.

kinematics (see for example Ehrhardt et al., 1971) with an
incident energy E = 250 eV, a scattering angle $\theta_A = 4°$ and an ejected
electron energy $E_B = 4$ eV. We see from Fig. 3 that all the second
Born features for the case of incident positrons are reversed with
respect to those corresponding to incident electrons. In particular,
the peaks are rotated in opposite directions, and the ratio of the
intensity of the forward peak to that of the backward peak is
increased for positrons while it is decreased for electrons in comparison with the first Born values.

Although differential measurements concerning positron impact
ionization are not yet available, total (integrated) positron ionization cross sections can be obtained. Thus, Griffith et al. (1979)
have deduced integrated cross sections for e^+-He ionization at intermediate energies. Using theoretical values of the elastic and excitation integrated cross sections, together with the measured total
(complete) cross sections of Twomey et al. (1977) and estimated
values of the positronium formation cross section (see Section 4),
Willis et al. (1981) have also deduced integrated e^+-He ionization
cross sections. They find, in agreement with Griffith et al. (1979)
that $\sigma_{ion}^+(e^+$-He$) \simeq \sigma_{ion}^-(e^-$-He$)$ to within 15% for incident energies
E > 60 eV. A similar result has been obtained for $e^\pm$-H ionization
by Ghosh, Basu and Mazumdar (1983), who find that the difference
between $\sigma_{ion}^+(e^+$-H$)$ and $\sigma_{ion}^-(e^-$-H$)$ is not significant for incident
energies E > 100 eV.

4. POSITRONIUM FORMATION

Positronium (Ps) formation, the process in which an incident
positron captures a bound electron to form the bound system (e^+e^-),
is a rearrangement collision of the same type as charge exchange.
The non-relativistic asymptotic behaviour of the cross section for
electron capture from atomic hydrogen by positrons with a very high
velocity $v \gg 1$ has been examined by Shakeshaft and Wadehra (1980).
They find that the leading contribution to the cross section is
provided by double scattering mechanisms, corresponding to the
second Born matrix elements of $V(pe^-)G_0^{(+)}V(e^+e^-)$ and $V(pe^+)G_0^{(+)}V(e^+e^-)$,
where $V(AB)$ is the interaction between particles A and B and $G_0^{(+)}$
is the "outgoing wave" (+) free Green's operator for three non-
interacting particles (Joachain, 1975). An interesting property
discussed by Shakeshaft and Wadehra is that for $v \gg 1$ the two
matrix elements interfere constructively or destructively according
to whether the orbital angular momentum quantum number ℓ of the final
positronium state is odd or even. As a consequence, the asymptotic v
dependence of the cross section for Ps formation is v^{-11} if ℓ is
odd and v^{-12} if ℓ is even.

For incident velocities which are relatively high (v > 1) but
not asymptotic, second and higher order calculations are very diffi-

cult to carry out. In this range of velocities, Ps formation has
been mostly studied by using first order theories, following the
pioneering work of Massey and Mohr (1954). In particular, the
process

$$e^+ + H(1s) \rightarrow Ps(1s) + H^+ \tag{6}$$

has been studied by Mandal and Guha (1979) using various first order
approximations, namely : the first Born approximation, a modified
Born approach, and the first order exchange approximation. More
elaborate calculations, using distorted waves, have been performed
by Mandal, Guha and Sil (1979), Shakeshaft and Wadehra (1980) and
Khan and Ghosh (1983 a, b). As an example, Table 5 shows the differential cross section for the process (6) at an incident positron
energy of 200 eV. The first Born (Jackson-Schiff) approximation is
compared to the distorted wave approximations of Shakeshaft and
Wadehra (1980) and Mandal, Guha and Sil (1979), in which static
distortion effects in the relative motion are included through
first and infinite order, respectively. It is clear from these
results that inclusion of distortion through first order is rather
satisfactory at small angles, but not at larger angles.

The integrated cross sections for the reaction (6), obtained
from the first Born approximation and from the above-mentionned
distorted wave treatments of Shakeshaft and Wadehra (1980) and
Mandal, Guha and Sil (1979) are compared in Table 6, for incident
positron energies of 50, 100 and 200 eV. The recent distorted wave
polarized orbital calculation of Khan and Ghosh (1983 a) give lower
cross sections ($0.27 \, \pi a_0^2$ at 54 eV) while the classical trajectory
Monte Carlo (CTMC) results obtained by Peach and Willis (1983) are
somewhat higher than those of Mandal et al.

Table 5. Differential cross section, in units of a_0^2/sr, as a function
of the scattering angle θ, for the reaction (6), at an
incident positron energy of 200 eV.

θ (deg)	FBA	SW (1)	MGS (2)
0	1.0(−1)	1.2(−1)	1.3(−1)
5	7.5(−2)	8.9(−2)	9.8(−2)
10	2.6(−2)	3.6(−2)	4.2(−2)
15	5.1(−3)	9.2(−3)	1.3(−2)
20	3.6(−4)	1.4(−3)	3.2(−3)
25	2.7(−5)	− 2.1(−4)	8.0(−4)

(1) Shakeshaft and Wadehra, 1980
(2) Mandal, Guha and Sil, 1979

Table 6. Total (integrated) cross sections (in units of πa_0^2) for the reaction $e^+ + H(1s) \rightarrow Ps(1s) + H^+$.

E (eV)	FBA	SW (1)	MGS (2)
50	4.6(-1)	4.3(-1)	5.1(-1)
100	4.5(-2)	4.8(-2)	5.7(-2)
200	2.4(-3)	3.0(-3)	3.8(-3)

(1) Shakeshaft and Wadehra, 1980
(2) Mandal, Guha and Sil, 1979

Mandal et al. (1979) and Peach and Willis (1983) have extended their calculations to analyze the reaction

$$e^+ + He(1\,^1S) \rightarrow Ps(1s) + He^+(1s) \qquad (7)$$

The CTMC results of Peach and Willis are again slightly higher than the distorted wave calculations of Mandal et al., both sets of results giving considerably larger integrated cross sections than those measured in the recent experiments of Charlton et al. (1983). Additional work is needed in order to understand the reasons of this discrepancy.*

Mandal et al. (1980) have also studied the process

$$e^+ + He(1\,^1S) \rightarrow Ps(ns) + He^+(1s) \qquad (8)$$

in the energy range 20-200 eV, using the first Born and the first order exchange approximations. Their results show that an n^{-3} law for the integrated cross section is satisfied in both approximations at each positron energy, the asymptotic limit of the formation cross section as $n \rightarrow \infty$ being reached within 0.1% beyond the excited state 16s.

Finally, we note that Ray et al. (1980) have applied the molecular Jackson-Schiff approximation to analyze the reaction

$$e^+ + H_2 \rightarrow Ps(ns) + H_2^+ \qquad (9)$$

in the energy range 50-1000 eV. They find that Ps formation involving the gerade transition of H_2^+ is the dominant process, and that the excited-state formation cross sections satisfy approximately an n^{-3} law.

---

*Note added in proof: The preliminary data obtained by L.S.Forneri, D.R.Cook and P.G.Coleman (this conference) are in much better agreement with the theoretical calculations of Mandal et al (1979) and Peach and Willis (1983).

REFERENCES

Blum K and Burke PG 1977 Phys. Rev. A16 163-8
Bransden BH and Hutt PK 1975 J. Phys. B : At. Mol. Phys. 8 603-11
Bransden BH and McDowell MRC 1969 J. Phys. B : At. Mol. Phys. 2
        1187-201
Byron FW Jr and Joachain CJ 1973 Phys. Rev. A8 1267-82
————— 1974 Phys. Lett. 49A 306-8
————— 1975 J. Phys. B : At. Mol. Phys. 8 L284-8
————— 1977 a Phys. Rep. 34 233-324
————— 1977 b Phys. Rev. A15 128-46
————— 1977 c J. Phys. B : At. Mol. Phys. 10 207-26
————— 1978 J. Phys. B : At. Mol. Phys. 11 2533-46
————— 1981 J. Phys. B : At. Mol. Phys. 14 2429-48
Byron FW Jr, de Heer FJ and Joachain CJ 1975 Phys. Rev. Lett. 35
        1147-50
Byron FW Jr, Joachain CJ and Piraux B 1980 J. Phys. B : At. Mol.
        Phys. 13 L673-6
Byron FW Jr, Joachain CJ and Potvliege R 1981 J. Phys. B : At. Mol.
        Phys. 14 L609-15
————— 1982 J. Phys. B : At. Mol. Phys. 15 3915-43
Charlton M, Griffith TC, Heyland GR and Wright GL 1980 J. Phys. B :
        At. Mol. Phys. 13 L353-6
Charlton M, Clark G, Griffith TC and Heyland GR 1983 J. Phys. B :
        At. Mol. Phys. (to be published)
Das JN and Biswas AK 1981 J. Phys. B : At. Mol. Phys. 14 1363-70
de Heer FJ, Wagenaar RW, Blaauw HJ and Tip A 1976 J. Phys. B :
        At. Mol. Phys. 9 L269-74
Ehrhardt H, Hesselbacher H, Jung K and Willmann K 1971 in Case
        Studies in Atomic Physics vol 2 (Amsterdam : North
        Holland) pp 159-208
Gerjuoy E and Krall NA 1960 Phys. Rev. 119 705-11
Ghosh AS, Basu M and Mazumdar PS 1983, this conference
Glauber RJ 1959 in Lectures in Theoretical Physics vol 1, ed WE
        Brittin (New York : Interscience) pp 315-414
Griffith TC, Heyland GR, Lines KS and Twomey TR 1979 J. Phys. B :
        At. Mol. Phys. 12 L747-53
Hutt PK, Islam MM, Rabheru A and McDowell MRC 1976 J. Phys. B :
        At. Mol. Phys. 9 2447-60
Joachain CJ 1975 Quantum Collision Theory (Amsterdam : North
        Holland)
Joachain CJ and Quigg C 1974 Rev. Mod. Phys. 46 279-324
Joachain CJ, Vanderpoorten R, Winters KH and Byron FW Jr 1977
        J. Phys. B : At. Mol. Phys. 10 227-38
Kauppila WE, Stein TS, Smart JH, Dababneh MS, Ho YK, Downing JP
        and Pol V 1981 Phys. Rev. A24 725-42
Khan P and Ghosh AS 1983 a Phys. Rev. A27 1904-9
————— 1983 b, this conference
Mandal P and Guha S 1979 J. Phys. B : At. Mol. Phys. 12 1603-11

Mandal P, Guha S and Sil NC 1979 J. Phys. B : At. Mol. Phys. $\underline{12}$ 2913-24

_____ 1980 Phys. Rev. A22 2623-9

Massey HSW and Mohr CBO 1954 Proc. Phys. Soc. A67 695-704

Morgan LA 1982 J. Phys. B : At. Mol. Phys. $\underline{15}$ L25-9

Mukherjee A and Sural DP 1982 J. Phys. B : At. Mol. Phys. $\underline{15}$ 1121-30

Parcell LA, McEachran RP and Stauffer AD 1983, this conference

Peach G and Willis SL 1983, this conference

Ray A, Ray PP and Saha BC 1980 J. Phys. B : At. Mol. Phys. $\underline{13}$ 4509-13

Scott T and Bransden BH 1981 J. Phys. B : At. Mol. Phys. $\underline{14}$ 2277-89

Shakeshaft R and Wadehra JM 1980 Phys. Rev. A22 968-78

Srivastava MK and Pathak A 1981 J. Phys. B : At. Mol. Phys. $\underline{14}$ L579-81

Stein TS and Kauppila WE 1982 Adv. At. Mol. Phys. $\underline{18}$ 53-96

Tip A 1977 a J. Phys. B : At. Mol. Phys. $\underline{10}$ L11-6

_____ 1977 b J. Phys. B : At. Mol. Phys. $\underline{10}$ L595-7

Twomey TR, Griffith TC and Heyland GR 1977 Proc. 10th Int. Conf. on Physics of Electronic and Atomic Collisions (Paris, Commissariat à l'Energie Atomique) Abstracts p 808

Wallace SJ 1973 Ann. Phys., NY $\underline{78}$ 190-257

Willis SL, Hata J. McDowell MRC, Joachain CJ and Byron FW Jr 1981 J. Phys. B : At. Mol. Phys. $\underline{14}$ 2687-704

Winick JR and Reinhardt WP 1978 Phys. Rev. A18 925-34

Winters KH, Clark CD, Bransden BH and Coleman JP 1974 J. Phys. B : At. Mol. Phys. $\underline{7}$ 788-98

POSITRONIUM FORMATION CROSS-SECTIONS IN VARIOUS GASES

T. C. Griffith

Dept. of Physics and Astronomy
University College London
Gower Street, London WC1E 6BT

Measurement of positronium formation cross-sections in a number of atomic and molecular gases has revealed some very interesting and unexpected behaviour as a function of positron energy.

1. INTRODUCTION

The emphasis, previously directed towards accurate measurements of total scattering cross-sections for positrons in different gases, has recently shifted to the problem of determining the scattering cross-sections for the various inelastic channels available for the positrons at different energies. The sharp rise in the total cross-sections from the positronium formation threshold, E_{Ps}, for the inert gases has been used by a number of authors to estimate the positronium formation cross-section, σ_{Ps}, at the energy, E_{ex}, of the first electronic excitation in a particular gas. It is assumed for this purpose that the elastic cross-section, σ_{el}, can be linearly extrapolated from below E_{Ps} up to E_{ex}. The results for these extrapolations have been discussed by Kauppila and Stein (1) and the observations of Charlton et al (2) have shown that at energies above E_{Ps} the values of σ_{Ps} do indeed increase sharply to reach a maximum at a few electron volts above threshold and then decrease fairly rapidly at higher energies.

Whilst the positronium formation channel has undoubtedly a unique character the excitation and ionization processes involving positrons are also of great importance especially in the energy

region where they compete with positronium formation. An attempt
to unravel the partial cross-sections at energies between 100 and
500 eV for positrons in helium was first made by Griffith et al (3)
who used a time-of-flight system to examine the energy distribution
of the inelastically scattered positrons. They arrived at the conclusion that, at the energies under consideration, ionization
appeared to be the dominant channel. More recently Coleman and
Hutton (4) and Coleman et al (5) have used a similar technique to
study the excitation process at energies extending from 2 to 10 eV
above E_{ex} in helium, neon and argon. Values of the ionization cross-
sections were also deduced for a limited range of energies up to
about 50 eV. It is noted that most of these cross-sections have
relatively small magnitudes and that when their values are added to
the sum of the extrapolated elastic cross-sections and the normalised
values of σ_{Ps} deduced from the data of Charlton et al (2), at
corresponding energies, the total sum is about 50% less than the
measured total cross-sections discussed by Kauppila and Stein (1).
It has also been shown by Cook et al (6) that a time-of-flight
arrangement can be used to estimate σ_{Ps} by measuring the number of
positrons which are totally removed from the beam by collisions in
various gases.

The experiments referred to above are all based on the results
obtained by attenuation of beams of positrons of well defined
energies in various gases. Information on the formation of positronium can also be obtained in a totally different type of
experiment namely, from observations on the lifetime spectra of
positrons in mixtures of gases at various densities. Observations
are made of the increase in the positronium fraction F, defined by
Heyland et al (7), as a function of small concentrations, C, of a
test gas in helium as the base gas. The increase in F is related
to both the mean momentum transfer cross-sections for helium,
obtained by extrapolation from the phase shifts of Campeanu and
Humberston (8) and to σ_{Ps} for the test gas at energies in the
range $E_{Ps} < E < E_{ion}$ where E_{ion} is the ionization threshold energy.
Values of σ_{Ps} at E_{ion} for all the inert gases (except helium) and
several molecular gases, obtained using this method, were reported
by Charlton et al (9). Despite the various assumptions involved
in deducing σ_{Ps} the values obtained from the total cross section
data at E_{ex} are remarkably consistent with those evaluated at E_{ion}
using the lifetime data. It is also worth noting that in the
lifetime studies the moderating positrons are involved in many
collisions with the gas atoms before forming positronium and that
subsequent collisions between energetic ground state positronium
(or any weakly bound excited state positronium that may have been
formed) and a gas atom may involve its dissociation if the kinetic

energy of the positronium exceeds its binding energy. This feature
is in sharp contrast to the conditions prevailing in a scattering
experiment where positrons of well defined energies traverse a
scattering cell containing gas at a density such that the number of
multiple collisions is kept as low as possible and collisions
between any positronium atoms formed in the interactions and other
gas atoms are also likely to be infrequent. Such conditions will
favour the survival and detection of energetic ground state and
excited state positronium atoms.

The preliminary results for the energy dependence of σ_{Ps}
reported by Charlton et al (2) were taken with a rather poor signal
to background ratio of only 1 to 4 and an energy range which only
extended for about 15 eV above the positronium formation threshold
in the four gases He, Ar, H_2 and CH_4 which were studied. These
measurements have now been repeated using a modified version of the
apparatus used for the earlier work. The energy range has been
extended to 150 eV and the signal to background ratio, at 5:1, has
been considerably improved. The new results, obtained with this
system, have revealed a number of unexpected features which constitute a severe challenge to the present state of the theory.

2. METHOD AND EXPERIMENTAL PROCEDURE

A brief account of the experimental arrangement used for the
present work was presented by Griffith et al (10). In order to
determine the cross-sections for positronium formation the earlier
system used by Charlton et al (2) has been extended and modified.
The main features of the apparatus consist of a 4 mCi ^{22}Na source
followed by a converter in the form of a vane system made from heat
treated tungsten and an accelerating grid which gave a beam of
about 1000 positrons per second at any desired energy between 3 and
150 eV with an energy spread of about ± 1.0 eV. The positron beam
from the moderator was transported by an axial magnetic field along
a 2 m long evacuated flight tube passing around a 15° bend to a
differentially pumped gas cell and thereafter to a channeltron
electron multiplier and retarding grid detection assembly. Gas can
be continuously leaked through the scattering cell via an auto-
matically controlled leak valve and the pressure is measured using
a Bevatron capacitance manometer. Three large Na I(Tℓ) counters
in triple coincidence were arranged symmetrically in the transverse
plane around the gas cell to detect the three γ's from the decay of
ortho-positronium formed in positron interactions in the gas. The
counters were shielded from the source and, where possible, from
one another using lead blocks and a lead collimator of length
150 mm with a 6 mm diameter hole was located in the flight tube
close to the converter. Up to 20 triple coincidences per minute
have been recorded with this arrangement at a signal to background
ratio which was 20 times better than for the earlier system of
Charlton et al (2).

Other improvements of the system have included a time-of-flight facility which is incorporated in the apparatus by replacing the strong source by a weak source together with its attendant scintillator and light guide placed at exactly the same location relative to the moderator. The time-of-flight start and stop pulses were derived in the usual way from the source scintillator and from the channeltron at the end of the flight path. The time-of-flight accessory was used for checking the alignment of the coils which provide the axial magnetic field of the beam transport system and also for determination of the mean energy of the positron beam; it has also been used, in a subsidiary experiment, for the determination of k, the gas cell scattering constant. When the strong source is used the beam intensity as a function of energy is monitored using the channeltron and its retarding potential element.

A data recording cycle comprised a determination of the triples counting rates at different energies with gas flowing through the gas cell followed in each case by a measurement of the triples background counting rate when the entire flight path was evacuated. The difference between these counting rates gave the number of coincidence signal events, T, per unit time. A correction had to be applied to T to allow for a background which arose from positrons which, under gas flow conditions, were scattered (without forming positronium) such that they hit the input or exit apertures. A small fraction of these scattered positrons were found to generate triples at a rate proportional to the attenuation of the beam. The magnitude of this background was assessed by investigating the behaviour of the triples counting rates at energies less than E_{Ps}. At low beam attenutations the positronium formation cross-section is given by

$$\sigma_{Ps} = \frac{4}{3} \ \frac{T}{n_o I_o} \left(\frac{\epsilon_2 \cdot k}{\epsilon_1 \ \ell_o} \right)$$

where I_o is the channeltron counting rate in vacuum, ϵ_1 the triple coincidence detection efficiency and ϵ_2 the detection efficiency of the channeltron for low energy positrons. As discussed by Charlton et al (11) the gas cell constant k is defined in terms of, n_o, the number density at the centre of the gas cell of geometric length ℓ_o and was determined in the present experiment using the time-of-flight system with electrons scattered in helium, neon and argon. The cross-sections determined in various other experiments in different laboratories were used for normalization and an average value for k of 1.18 ± 0.02 was obtained in this way. Relative cross-sections can be obtained directly by evaluation of $(T/n_o \ell_o)$ but absolute cross-sections are only known if

$(\varepsilon_2/\varepsilon_1)(k/\ell_o)$ has also been determined. Both ε_1 and ε_2 can, in principle, be found either by direct measurement or by calculation but, so far, this has not been accomplished with sufficient accuracy. Absolute values of the cross-sections have, therefore, been deduced by a different normalisation procedure based on estimating the values of σ_{Ps} in the energy interval $E_{Ps} < E < E_{ex}$ from the total cross-sections for e^+-argon reported by Charlton et al (12) and Kauppila et al (13) and, as mentioned in §1, it has to be assumed that the elastic cross-section can be extrapolated linearly up to E_{ex}. The observed value of T at an energy of 11.5 eV in argon was compared with the value of σ_{o-Ps} deduced from the total cross-section measurements and a mean value of $\varepsilon_2/\varepsilon_1 = 0.006 \pm 0.001$ was obtained. Uncertainties in the extrapolation and in the magnitude of the total cross-section at 11.4 eV could contribute up to 10% as a systematic error on the values of σ_{Ps}.

A number of other factors that may have contributed to systematic errors in the earlier experiment of Charlton et al (2) have also been investigated. One test involved monitoring the triples counting rates in argon as a function of positron energy for different lengths and diameters of the gas cell (amounting to a change in the surface to volume ratio of 20:1 for the gas cell) and also the effect of coating the inside walls of the gas cell with fine MgO powder. It was concluded that, at positron energies less than 40 eV, the variation of T due to premature 2γ annihilation of ortho-positronium on the walls of the gas cell, was negligible. Similar tests were performed to investigate the effect of changing the diameters of the entrance and exit apertures of the gas cell where values of 4 mm, 6 mm and 8 mm respectively were used. On changing the diameters from 6 mm to 4 mm at positron energies near 35 eV the values of T were not observed to change by an amount greater than 5%, thus confirming that the escape of fast ortho-positronium from the gas cell before decaying was not a serious source of error.

Positrons which have undergone a large angle elastic collision in the gas cell may, due to spiralling, have a much longer path length for a further collision involving positronium formation. In this case the observed triples rates will be exaggerated. The gas density for which T was determined was maintained at a level such that less than 15% of the incident beam was attenuated and the probability of multiple collisions was at an acceptably low value. The triples rates as a function of gas pressure were investigated at different energies for all gases and no departure from linearity (which would have implied multiple collision effects) could be detected in the pressure region used for the determination of σ_{Ps}.

The various sources of error discussed above could mean that the cited values of σ_{Ps} may be in error on an absolute scale by as much as 15 to 20%. It is estimated that any distortion of the σ_{Ps} versus energy curves at energies between 20 and 30 eV due to the combined effect of escaping fast positronium, premature annihilation of ortho-positronium on the walls and uncertainties in the background subtractions, is less than 10%.

3. RESULTS AND DISCUSSION

Each of the five inert gases, He, Ne, Ar, Kr and Xe and the molecular gases H_2, N_2, CO_2, CH_4 and O_2 have been examined in detail at energies between 3 and 150 eV. The variation of σ_{Ps} with the mean energy show quite distinctive features which have not previously been suggested either experimentally or theoretically. The curves for the inert gases are shown in Figure 1 and for the molecular gases in Figure 2. Each gas exhibits a different curve but it should be noted that for seven out of the ten gases investigated there appears to be a prominent maximum in the cross-section at a few electron volts above the ionisation threshold. This is a structure that exists in addition to the large peak corresponding to the formation of ground state positronium which, in all cases, rises rapidly from the positronium formation threshold. It is only for the case of helium that the secondary peak above the ionisation threshold is larger than the peak starting from E_{Ps}. In CO_2 there appears to be only one rather broad peak.

It should be recalled that the incident positron beam has an energy width of ± 1.0 eV. The exact way in which this spread influences the shape of the cross-section curves has not yet been analysed in detail, It is clear that it has the effect of broadening the peaks and that corrections should lead to sharper peaks and to a sharper rise of the cross-sections from the threshold at E_{Ps}. Bearing in mind that the present data are much more accurate and detailed it is interesting to note that over the rather limited energy range of the earlier measurements of Charlton et al (2) there is quite good agreement between the two sets of data for argon, H_2 and CH_4. In the case of helium, however, the present cross-sections are appreciably higher than the earlier ones at the higher energies and the discrepancy is probably due to a systematic error caused by poor background subtraction in the earlier work.

One further feature of the results presented here is that on extending the energy range of the positrons up to 150 eV it is seen that some positronium can be formed at energies above 100 eV. It is also to be noted that, using the $e^+ - $ Ar data for normalisation, the values of σ_{Ps} for xenon are appreciably larger than those which can be estimated from the total cross-sections of Dababneh

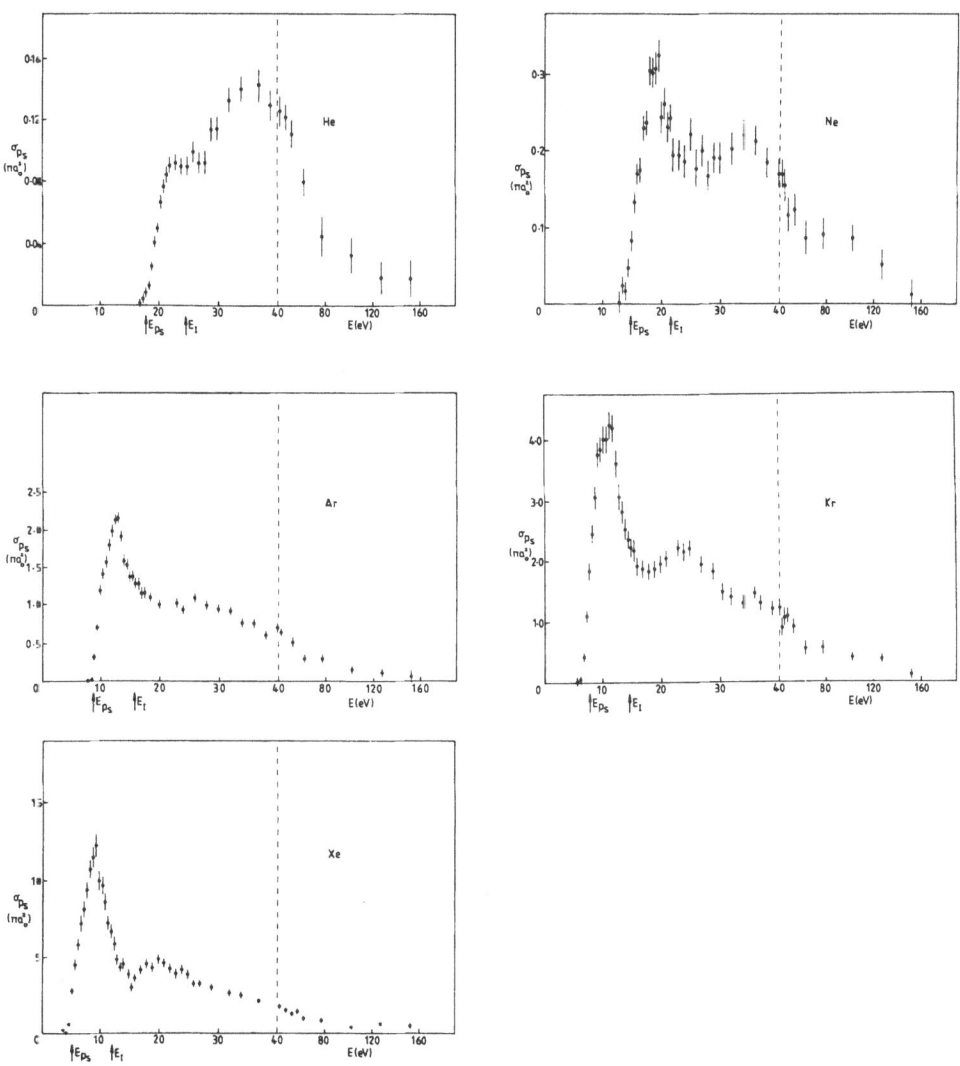

Figure 1: Positronium formation cross-sections, σ_{Ps}, as a function of the mean energy of the positrons for helium, neon, argon, krypton and xenon.

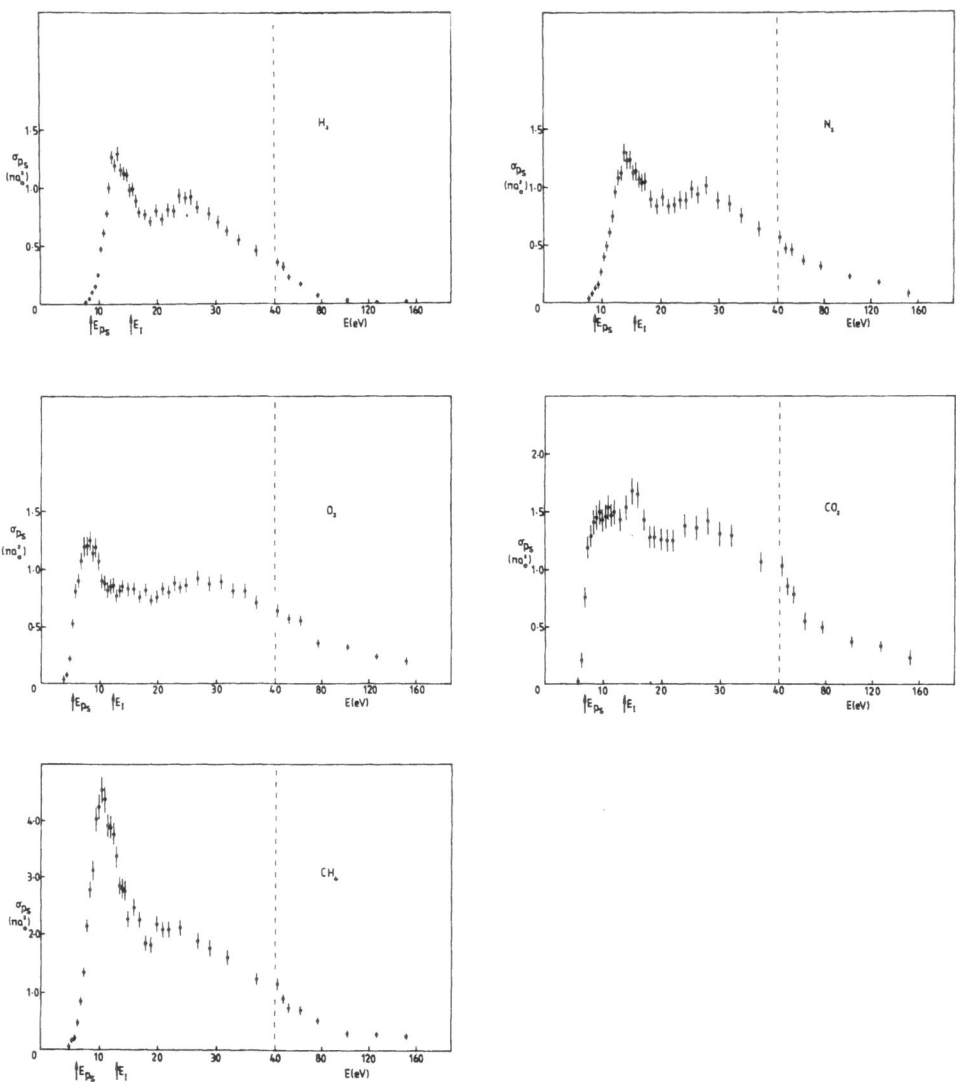

Figure 2: Positronium formation cross-sections, σ_{Ps} as a function of the mean energy of the positrons for H_2, N_2, O_2, CO_2 and CH_4.

et al (14). The e^+ - Xe total cross-section measurements of
Dababneh et al (14), Coleman et al (15) and Sinapius et al (16) are,
however, not in good agreement below E_{Ps} so that σ_{Ps} for Xe cannot
easily be estimated from these data. Difficulties of this nature are
not encountered with the helium, neon or krypton results.

     The observed structure in the measured cross-sections pose a
severe challenge to the present state of the theory. Humberston
(17) has reviewed the theoretical aspects of positronium formation
and has pointed out that even for helium the predictions for σ_{Ps},
using different theoretical approaches, were widely different.
The only calculations for the heavier inert gases are those of
Gillespie and Thompson (18) for neon and argon. All these calculations are in poor agreement with the experimental results and at
best only apply to the primary peak which is assumed to represent
formation of positronium in the ground state. The secondary peak
may be associated with the ionisation of the atom or molecule by
the incoming positron and Bransden (19) and Klar (20) have both
suggested that for ionisation by positrons at energies close to
the threshold, E_{ion}, it is conceivable that there will be strong
correlation between the positron and the electron involved in the
collision. Bransden (19) has further suggested that this might
mean that positronium formation in a continuum state is an important
process. The present results would appear to support this conjecture and should provide an impetus for renewed theoretical activity
on this problem. Experimentally it is conceivable that a process of
this kind may lead to the release of low energy photons from excited
states of positronium and work is in progress to incorporate a
photon detector as part of the scattering cell of the system used
for the work described here. It is appropriate to note that in
helium at 30 eV, σ_{Ps} amounts to roughly 25% of the total e^+- helium
cross-sections so that any theoretical estimates of the excitation
or ionisation cross-sections at these energies must take full
account of positronium formation. In some of the molecular gases
the values of σ_{Ps} are relatively low. In nitrogen at 25 eV, for
example, σ_{Ps} is only about 10% of the total cross-section. If the
new values of σ_{Ps} reported here for helium, neon and argon are added
to the extrapolated elastic cross-sections and the measured values
of the excitation and ionisation cross-sections reported by Coleman
et al (5) they give total cross-sections for positron interactions
which are substantially lower than those discussed by Kauppila and
Stein (1).

Addendum: At this Workshop P.G. Coleman reported new results for
positronium formation cross-sections in He and Ar which are in serious
disagreement with the present values. Several tests and modifications of the apparatus used for the present work have been initiated
in an attempt to resolve these differences.

Acknowledgements

The work described in this article has been funded by the SERC and has been carried out in collaboration with Dr. M. Charlton, Dr. GR Heyland and Mr. G. Clark.

References

1: WE Kauppila and TS Stein 'Positron-gas cross-section measurements' Can.J. Phys. 60:471 (1982)

2: M Charlton, TC Griffith, GR Heyland, KS Lines and GL Wright 'The energy dependences of positronium formation in gases' J. Phys. B 13:L 757 (1980)

3: TC Griffith, GR Heyland, KS Lines and TR Twomey "Inelastic scattering of positrons by helium atoms at intermediate energies' J. Phys. B 12:L 747 (1979)

4: PG Coleman and JT Hutton 'Excitation of Helium Atoms by Positron Impact' Phys. Rev. Lett. 45:2017 (1980)

5: PG Coleman, JT Hutton, DR Cook and CA Chandler 'Inelastic scattering of slow positrons by helium, neon and argon atoms' Can. J. Phys. 60:584 (1982)

6: DR Cook, PG Coleman, LM Diana and SC Sharma 'Positronium formation in helium' Proc. 6th Int. Conf. on positron annihilation, Univ. Texas at Arlington (1982) p.87

7: GR Heyland, M Charlton, TC Griffith and GL Wright 'Positron lifetime spectra for gases' Can. J. Phys. 60:503 (1982)

8: RI Campeanu and JW Humberston. The scattering of s-wave positrons by helium. J. Phys. B 10:L 153 (1977)

9: M Charlton, TC Griffith, GR Heyland and KS Lines "Cross-section for Positronium formation in gases' J. Phys. B. 12:L 633 (1979)

10: TC Griffith, M Charlton, G Clark, GR Heyland and GL Wright 'Positrons in gases - a Progress Report'. Proc. 6th Int. Conf. on positron annihilation, Univ. Texas at Arlington p.61 (1983)

11: M Charlton, TC Griffith, GR Heyland and GL Wright ' Total scattering cross-sections for low energy positrons in the molecular gases H_2, N_2, CO_2, O_2 and CH_4' J. Phys. B 16:323(1983)

12: M Charlton, TC Griffith, GR Heyland and GL Wright. Private communication (1983) - to be published.

13: WE Kauppila, TS Stein and G Jesion. Direct observation of a Ramsaur-Townsend Effect in Positron-Argon Collisions' Phys. Rev. Lett. 36:580 (1976)

14: MS Dababneh, WE Kauppila, JP Downing, F Lapperriere, V Pol, JH Smart and TS Stein "Measurement of Total Scattering Cross-Sections for Low Energy Positrons add Electrons Colliding with Krypton and Xenon' Phys. Rev. A22:1872 (1980)

15: PG Coleman, JD McNutt, LM Diana and JT Hutton 'Measurements of total cross-sections for the scattering of positrons by argon and xenon atoms' Phys. Rev. A22:2290 (1980)

16: G Sinapius, W Raith and WG Wilson.'Scattering of low energy positrons from noble-gas atoms' J. Phys. B 13:4079 (1980)

17: J. Humberston. "Theoretical aspects of positron collisions in gases' Adv. in Atomic and Molecular Phys. 15:101 (1979)

18: ES Gillespie and DG Thompson 'Positronium formation in neon and argon' J. Phys. B 10:3543 (1971)

19: BH Bransden. "Comments in round table discussion at Int. Conf. on positron scattering in gases' Can. J. Phys. 60:567 (1982)

20: H Klar 'Threshold ionisation of atoms by positrons' J. Phys. B 19:4165 (1981)

POSITRONIUM: RECENT FUNDAMENTAL AND APPLIED RESEARCH

D. W. Gidley*   P. G. Coleman

Bell Laboratories   Department of Physics
Murray Hill   The University of Texas
New Jersey 07974   at Arlington
U.S.A.   Arlington, Texas 76019
                         U.S.A.

INTRODUCTION

In the first workshop on positron scattering in gases the status of positronium (Ps)-related tests of quantum electrodynamics was summarized.[1] In the ensuing 2 years there has been steady progress in this area.[2] However, and more in keeping with the emphasis of intense positron beams in the present workshop, we have expanded our topic to include fundamental experiments in which Ps is principally used to probe matter, and the various methods used to form Ps, Ps* (excited state Ps), and Ps⁻ (the Ps negative ion). Of course, a complete summary of these fields of research is beyond the scope of this paper. Thus, we have limited ourselves to highlighting selected experiments that we hope will indicate the diversity of research that involves Ps. This paper is organized into 3 sections: methods of forming Ps, Ps formation as a probe of matter, and fundamental properties of the Ps atom. For recent reviews of Ps see also references 3 and 4.

FORMATION OF Ps, Ps*, AND Ps⁻

Ps Formation in Gases

In typical positron lifetime experiments in gases, beta

---

*On leave of absence from The University of Michigan, Ann Arbor, Michigan 48109.

positrons slowing down through collisions with gas molecules can form Ps with a high probability. The Ore model for Ps formation is based upon the idea that if a positron is thrown into the energy region $(E_i-6.8eV)<E<E_{ex}$ (the "Ore Gap"), where E_i and E_{ex} are the threshold energies for ionization and excitation, respectively, then it will form Ps. The upper limit is extended to E_i if atomic excitation does not compete with Ps formation below E_i. Above E_i, the model assumes that either (a) excitation and ionization swamp Ps formation, or (b) the energetic Ps, if formed, is immediately dissociated in a subsequent collision. For the lighter noble gases the fraction of positrons which form Ps lie within the Ore Gap limits, but in Kr and Xe the fraction is low.[5] In many molecular gases the Ps fraction (F) is much higher than predicted by the Ore model, and is strongly density dependent (e.g., $F \approx 0.8$ for ethane at 150 amagat[6]). Griffith et al.[7] suggest that the low F values for Kr and Xe may be due to the formation of Ps-atom complexes which decay promptly, and the high F values for high-density gases have been discussed in terms of the spur model.[8] It is apparent that a full understanding of the interesting density dependences of F in gases is yet to be gained.

Ps formation in low-density gases (i.e., in single positron-molecule collisions) is currently being studied experimentally in London[9] and Arlington[10] and theoretically by a number of researchers.[11,12] Recent advances in this area are reviewed by Professor Griffith and we shall not dwell on them here, except to point out that the Ore Gap picture has to be abandoned for single positron-atom interactions. The threshold for Ps formation is still, of course, $E_i-6.8eV$, but there is no distinct upper limit to the energy region in which Ps may be formed, since neither assumption (a) nor (b) above is necessarily valid.

Ps Formation in Powders

The formation of Ps with high efficiency by positrons incident on small (diameter < 100Å)-grained powders of MgO, SiO$_2$, and Al$_2$O$_3$ was observed 15 years ago.[13] It was the paper by Curry and Schawlow[14] on the emission of free Ps into the vacuum from MgO which led to the development of the successful MgO moderator for slow positron beam production at University College London.[15] Ps formed in such grains and migrating into the vacuum between the grains does not reenter a grain;[16] its lifetime is, therefore, reduced only minimally by pick-off at the grain surface, by an amount dependent upon grain density and size.[17] This observation is in contrast to those of Ps formed in the pores of silica gels[18] and in Vycor glass,[19] where it appears that Ps is localized in an attractive well close to the surface of the pores.

Ps Formation on Solid Surfaces

The formation of Ps with high efficiency at solid surfaces bombarded with slow positrons was first observed by Canter, Mills and Berko.[20] Mills[21] confirmed that Ps formation, forbidden in bulk metals, occurs following the diffusion of implanted slow positrons back to the surface. Thermal and non-thermal Ps has been observed from surfaces,[22] the former resulting from thermal desorption of positrons bound in surface states. Further studies on Ps emission from surfaces are described in several recent publications,[23,24,25] and the references therein.

The production of Ps beams of thermal energies or a few eV is feasible with the advent of high-intensity slow positron beams; the production of a beam of controllable-energy Ps by photoionizing the Ps negative ion has been suggested by Mills.[26]

Sources of Cryogenic Ps

The fact that thermal Ps is formed in the thermal desorption of a positron from the surface image well suggests that cryogenic Ps might be produced on some appropriate surface. Production of very low velocity Ps is important in reducing second order Doppler shifts in the 1^3S_1-2^3S_1 2 photon experiments and in reducing wall collisions in the triplet decay rate measurements (at room temperature triplet Ps has an annihilation length of only 1 cm). Both Lynn[27] and Mills[28] have observed thermal desorption of Ps at room temperature using an Al(111) target after an exposure to O_2 of about 500 Langmuirs. The Michigan group has observed the same effect using a half monolayer of Cs on Ni(110). The activation energy, E_a, for the desorption process is

$$E_a = \phi_- + B - 6.8eV,$$

where B is the binding energy of the positrons on the surface and ϕ_- is the electron workfunction. The effect of 0 or Cs adsorption is to lower E_a by reducing ϕ_- and thus decreasing the desorption temperature to room temperature or below.

Ps* Formation

The formation of Ps in vacuum on solid surfaces using a low energy positron beam[20] led to the first observation[29] of Lyman-α radiation from Ps*. In an effort to improve the relatively low yield of Ps* (10^{-3} per incident positron) obtained on a Cu target in moderate vacuum the Brandeis group has investigated W and Cu surfaces in ultra high vacuum.[30] The highest yield observed on polycrystalline W was 4×10^{-3} at an incident beam energy of 10eV. No sensitivity to surface contamination was observed. Since the Ps* workfunction for the nth Rydberg level,

$$\phi_{Ps}(n) = \phi_+ + \phi_- - \frac{6.8}{n^2} \, ,$$

is almost certainly positive for all metals, the formation process
must involve non-thermal positrons and presumably a charge exchange
mechanism at the surface.[30]

Ps Spin State Selection

We note briefly that spin polarized positron beams could allow
the selection of certain spin states of Ps. Triplet states with
m = +1 or with m = -1 could be selected by forming Ps with a
polarized beam in a strong magnetic field. The magnetic field
might be eliminated by using a magnetized target (so that the
captured electron is also spin polarized).[31] The Michigan group[32]
has produced low energy positron beams with polarization as high
as 70%.

Formation of Ps⁻

The first observation of the Ps negative ion was reported by
Mills[26] in 1981. The apparatus used is shown in Fig. 1 (a). Slow
positrons impinge upon a thin carbon film; approximately 1 positron
in 10^4 picks up two electrons and emerge as Ps⁻ ions which are
accelerated by an electrostatic field. The signature of the Ps⁻
ions is a blue-shifted annihilation line, measured by a Ge(Li)
detector (Fig. 1 (b)).

Ps FORMATION AS A PROBE OF MATTER

In this section we present selected experiments that demonstrate
the wide range of applicability of Ps as a probe of matter. These
areas of research include astrophysics, condensed matter and surface
physics, biophysics, and molecular physics.

Galactic Ps

One of the most intriguing examples of probing matter with
positrons involves the detection of a narrow 511 keV annihilation
line from the direction of the galactic center. This line was
observed on some 7 different occasions from 1971 to 1980 but within
one year it disappeared.[34,35,36] Thus this mysterious object must
be compact (less than 1 light year in diameter) and a very intense
positron source (10^{43} positrons/sec if isotropic). In addition,
the role of Ps formation and the subsequent decay of the singlet
state as the predominant source of 511 keV radiation is not yet
clear. Should the galactic center positrons be annihilating in a
cold, low density gas of hydrogen then virtually all positrons
should annihilate from the bound state of Ps.[37] This is due to the
large charge exchange cross section shown in Fig. 2 that produces

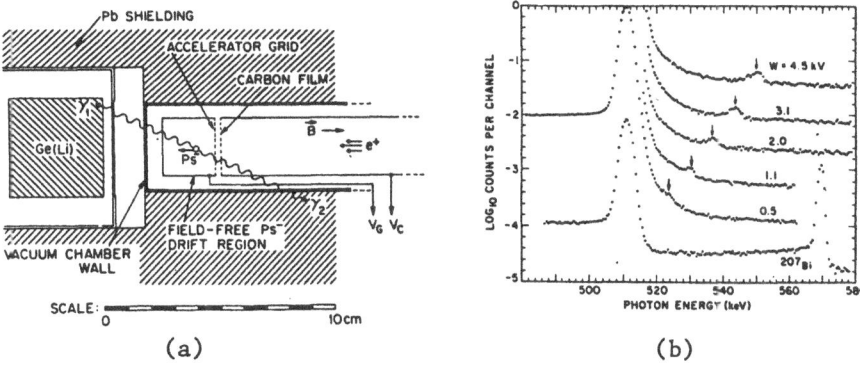

Fig. 1. (a) Apparatus used by Mills (ref. 26) to observe the
         formation of Ps⁻. (b) Ge(Li) spectra showing the blue-
         shifted annihilation line due to the decay of Ps⁻
         accelerated by 0.5-4.5 kV. From ref. 26.

Ps with relatively high energies (of order 30 eV). As a result a
large continuum of 3 γ annihilation should be observed along with
a Doppler broadened 511 keV line with FWHM of 5-6 keV.[37] At this
time there is no conclusive evidence for the 3 γ continuum (which
is difficult to differentiate from the background continuum). The
linewidths at 511 keV for the two high resolution measurements[33,34]
are consistent with the Ge detector resolution of about 3 keV. Thus
it is deduced[37] that the annihilation medium must be less than
50,000 K and at least 10% partially ionized. The effect of the
ionized medium is to quickly slow the positrons down to below Ps
formation threshold, thus avoiding electron capture to fast Ps.
Expected linewidths are then consistent with the observations. Dr.
Drachman will consider this topic in more detail in his paper.

     An experiment to test annihilation linewidth predictions in a
controlled environment is now underway at Bell Labs by Leventhal,
Brown, Mills, and Gidley. A 200 eV beam of positrons is magnetically guided into a one meter long target chamber filled with H_2
gas. Electrostatic grids and a 1.5 kG axial magnetic field trap
those positrons that make at least one collision with a gas
molecule. The gas pressure is kept low enough ($<10^{-3}$ torr) so that
Ps can decay in flight without significant probability of colliding
with the gas. A Ge(Li) detector with 1.4 keV resolution at 511 keV
will monitor annihilation γ rays from the target region. The line
narrowing effect of free electrons can be checked by injecting
electrons into the target region. Preliminary results should be
available this summer.

Surface Defect Studies

     The measurement of the fraction of incident slow positrons

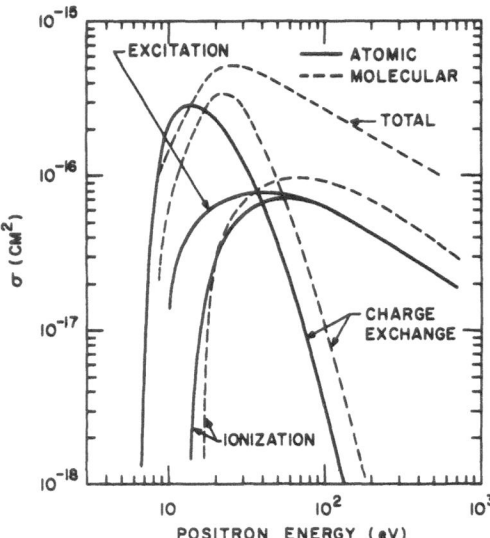

Fig. 2.   Theoretical cross sections for charge exchange, excitation,
          and ionization in atomic hydrogen (solid lines) and in
          molecular hydrogen (dashed lines) as a function of
          positron energy. The total cross section curve is an
          experimental result for molecular hydrogen. From ref. 37.

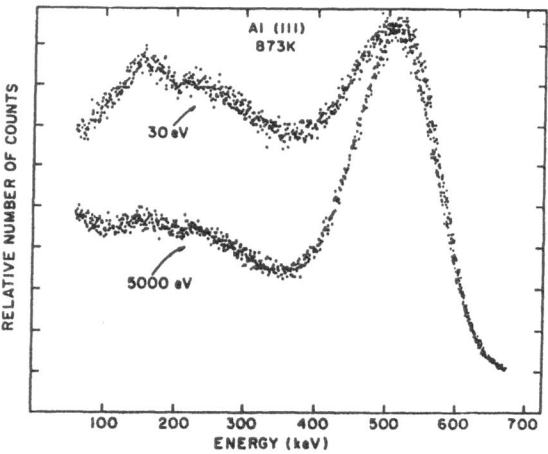

Fig. 3.   NaI spectra for 30 and 5000 eV positrons incident on
          Al(111) at 873K. From ref. 38.

which diffuse back to a surface to form Ps is a powerful method for studying near-surface defects, which can trap the diffusing positrons. Lynn and his coworkers at Brookhaven have employed this technique extensively. They obtain the Ps formation fraction by measuring the 511 keV photopeak-to-valley ratio using a NaI detector, as illustrated in Fig. 3. As more Ps is formed, the counts in the valley region increase at the expense of those in the photopeak, signifying a decrease in 2 γ decays and a corresponding increase in 3 γ (o-Ps) decays. An excellent overview of this work is presented by Lynn.[38] Vacancy formation enthalpies,[39] positron activation energies from surface states, specific trapping rates,[24] trapping in oxide overlayers,[40] and anomalous diffusion lengths[41] are among the quantities measured by this technique. Fig. 4 shows the Ps fraction F vs. temperature for different incident positron energies. The low-temperature increase in F is ascribed to thermal desorption of positrons from the surface state as Ps, and the high-temperature decrease in F is attributed to positron trapping in thermally-generated vacancies.

Surface Magnetism Studies

The fact that Ps formation can only occur in the low density tail of the electron distribution at the surface of a metal has been exploited by the Michigan-GMR group to investigate surface magnetism.[31] A spin polarized low energy beam of positrons has been used to measure the polarization, p_{e-}, of electrons captured at a Ni(110) surface that has been magnetized in the $(1\bar{1}1)$ direction. An asymmetry in the triplet Ps formation rate, A_T, is observed when either the Ni magnetization or the positron beam polarization is reversed. P_{e-} can then be deduced from

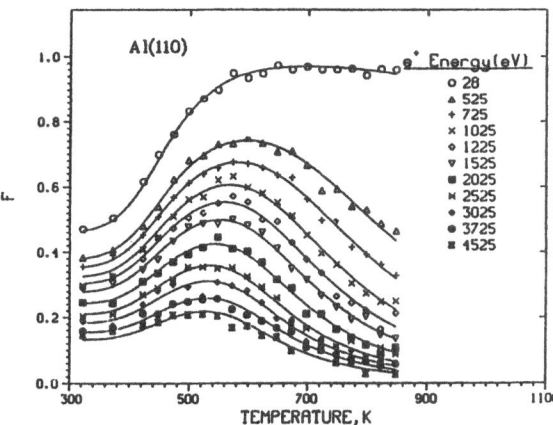

Fig. 4. Ps fraction F for 28-4525 eV positrons incident on Al(110) vs. temperature. From ref. 38.

$$A_T = \frac{1}{3}\,\vec{P}_{e^+} \cdot \vec{P}_{e^-}\ ,$$

where P_{e^+}, the positron beam polarization, is 0.50.   Experimental
results are shown in Fig. 5 and are explained in greater detail in
ref. 31.   To summarize the results, the surface magnetization is
found to differ markedly from that of the bulk in the temperature
range $0.46 < T/T_c < 1.0$ where T_c = 633 K, the Ni Curie temperature.
The shape of the curve is consistent with polarized low energy
electron scattering measurements and calculations of the surface
magnetism (see refs. in 31).   The asymmetry in triplet Ps formation is very sensitive to surface conditions, dropping a factor of
5 with either a half monolayer of 0 adsorbed on the surface or after
Ar ion sputtering of the surface.   Therefore this technique does
appear to be both spin sensitive and surface selective.   A quantitative understanding of A_T must incorporate the fact that $\phi_{Ps} \approx$
$-3eV$ for Ni and thus electron capture is allowed only within 3eV of
the Fermi level.   In an effort to untangle any energy and momentum
dependent features[31] measurements are now underway on Ni(100) and
Ni(111) surfaces.   An experiment to measure P_{e^-} for thermally
desorbed positrons from the surface image well is also in progress.

Studies of Optically Active Molecules

A second application of a polarized beam deals with a search
for a triplet Ps asymmetry in biologically relevant molecules.[42]
In dissymmetric (optically active) molecules the spin orbit interaction will produce a helicity in the bound molecular electrons.
Such an electron helicity would give rise to an asymmetry in
triplet Ps formation when low energy <u>helicitized</u> positrons are

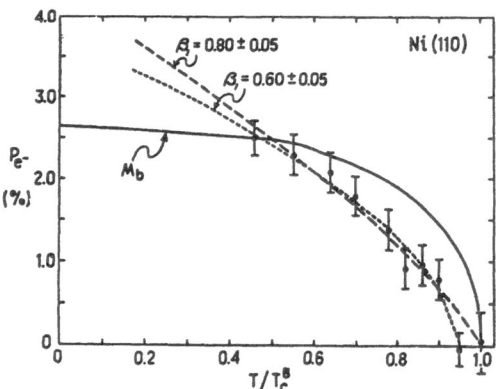

Fig. 5.   Temperature dependence of P_{e^-}. Dashed and dotted curves
          are fits of the data to $P_{e^-}(T) = P_{e^-}(0)[1-T/T_c]^{\beta_1}$ with T_c
          fixed at 633 K and with T_c unconstrained, respectively.
          The bulk magnetization, M_b, has been normalized to the
          datum point at $T/T_c^B = 0.46$. (From ref. 31).

incident on the target. The observation of $A_T \neq 0$, and thus an indication of bound electron helicity, necessarily leads to the conclusion[43] that preferential, isomer dependent ionization of optically active molecules by beta decay radiation will occur. Thus, this experiment probes a possible causal mechanism between the helicity of beta decay electrons and the almost complete dominance of L-amino acids. For details and preliminary results see references 42 and 43.

Density Fluctuations in Gases

Recently, Sharma et al.[44] have reported measurements of the o-Ps decay rate in ethane vs. gas density (see Fig. 6). The deviation of the decay rate from a linear dependence on density is shown to be correlated with the deviation of the real gas density from that of an ideal gas. It is attributed to the presence of density fluctuations in the gas, with o-Ps preferentially decaying in regions of lower-than-ambient density. These measurements constitute a departure from traditional Ps lifetime measurements in gases in that Ps decay is being used as a probe of intermolecular interactions, rather than to monitor the positron-molecule interaction itself.

FUNDAMENTAL PROPERTIES OF Ps

In this section we update new developments in measurements that focus on the properties of Ps as a test of quantum electrodynamics. We loosely divide this topic into 2 catagories: those experiments dealing with decay properties (decay rates, decay modes, γ ray energies and angular correlation); and those experiments involved with the spectroscopy of Ps.

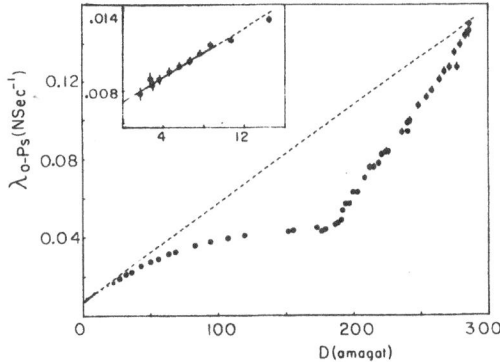

Fig. 6. o-Ps annihilation rate vs. ethane gas density at 306.4K. The dashed line is an extrapolation of the fit to the low-density data points (see inset). From ref. 45.

Ps Decay Rates

The annihilation decay rates of singlet, $^1S_0$, and triplet, $^3S_1$, ground state positronium are, respectively,

$$\lambda_S = \lambda_2 + \lambda_4 + \lambda_6 + \ldots$$

$$\lambda_T = \lambda_3 + \lambda_5 + \lambda_7 + \ldots$$

where the subscript denotes the number of photons in the final state. The total decay rate is the sum of the individual decay rates into each allowed final state. Charge conjugation invariance requires the singlet state to decay into an even number of photons and the triplet state to decay into an odd number. It should be noted that two experiments[45,46] have searched for the C-parity forbidden decays $\lambda_3(^1S_0)$ and $\lambda_4(^3S_1)$ respectively. In these experiments upper limits of 3×10^{-6} were set on the branching ratios $\lambda_3(^1S_0)/\lambda_2(^1S_0)$ and $\lambda_4(^3S_1)/\lambda_3(^3S_1)$. Although the forbidden decays are strongly suppressed, the present upper limits are barely sensitive to the expected branching ratios calculated with ad hoc C-violating modifications to the minimal coupling Hamiltonian.

Since the last workshop calculations of λ_4[47,48,76] and λ_5[48] have been reported. In both cases $\lambda_4/\lambda_2 \approx \lambda_5/\lambda_3 \approx 10^{-6}$ and can be neglected, as expected, where comparing theory and experiment. The theoretical value for λ_2 through the first order radiative correction was first calculated by Harris and Brown[49] to be

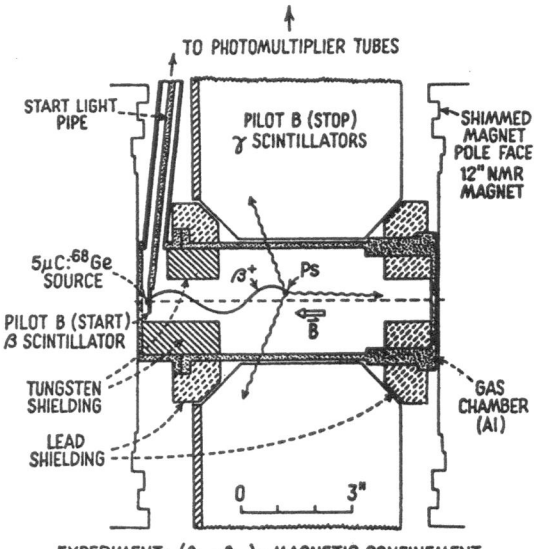

EXPERIMENT: $(\lambda_S - \lambda_T)$ - MAGNETIC CONFINEMENT

Fig. 7. The gas chamber and detector arrangement in which λ_S and λ_T are measured. From ref. 54.

$$\lambda_2(\text{theory}) = 1/2 \; \alpha^5 \; \frac{mc^2}{\hbar} \; (1 - \frac{\alpha}{\pi} \; (5 - \frac{\pi^2}{4}) = 7.9852 \; \text{nsec}^{-1}.$$

This result has been verified by Cung et al.,[50] Freeling,[51] and Tomozawa.[52]

The first experiment to rigorously test the 0.6% radiative corrections has been recently completed at Michigan.[53] Ps is formed in isobutane gas in a uniform magnetic field of about 4 kG as shown in Fig. 7. The magnetic field mixes the m=0 triplet and singlet states and, as a result, the m=0 triplet decay rate is increased to

$$\lambda_T' = \frac{1}{1+y^2} \; \lambda_T + \frac{y^2}{1+y^2} \; \lambda_S \; ,$$

where $\quad y = \dfrac{x}{1 + \sqrt{1+x^2}} \quad$, and $\quad x = \dfrac{2g'\mu_B B}{h\Delta\nu} \approx \dfrac{B(\text{kG})}{36.287}$.

Thus the annihilation lifetime spectrum has 2 exponential components: the unperturbed decay from the m = ± 1 states; and the "quenched" decay from the m = 0 state (at 4 kG the lifetime is about 30 nsec). Measurement of these decay rates, λ_T and λ_T', at gas pressures ranging from 200–1400 torr allows one, after extrapolation to zero gas density, to solve the above equation for λ_S. Measurements were made at 3 different magnetic fields and the averaged result is

$$\lambda_S = 7.994 \pm 0.011 \; \text{nsec}^{-1} \; ,$$

in agreement with the Harris and Brown calculation at the 0.15% level.

The above experiment also produced a new measurement of λ_T at comparable (slightly improved) accuracy to the previous results. The new result[53] is

$$\lambda_T = 7.051 \pm 0.005 \; \mu\text{sec}^{-1}.$$

The present theoretical value for the 3 photon decay rate is[54,77]

$$\lambda_3(\text{theory}) = \frac{2(\pi^2-9)}{9\pi} \; \frac{\alpha^6 mc^2}{\hbar} \; [1 - \frac{\alpha}{\pi} \; (10.266 \pm 0.011) - \frac{\alpha^2}{3}\ln\alpha^{-1}]$$

$$= 7.0386 \pm 0.0002 \; \mu\text{sec}^{-1} \; .$$

Current experimental values of λ_T are summarized in Table 1. All of the measurements are 1-2 standard deviations above theory. The uncalculated radiative correction terms of $\theta(\alpha^2)$ would require a coefficient of about 25 to account for the difference. The increase in λ_T(theory) due to λ_5 is now known to be negligible.[48] Some of the self energy terms in λ_3(theory) have also been verified

Table 1

Current experimental values for the triplet Ps decay rate, λ_T.

Medium	Ref.	$\lambda_T(\mu sec^{-1})$	
Gas	(55)	7.056	0.007
Gas	(56)	7.045	0.006
Gas(B field)	(53)	7.051	0.005
Vacuum	(57)	7.050	0.013
Powder	(55)	7.067	0.021

analytically.[58] On the other hand, most of the experimental
systematic errors tend to increase the measured decay rate. Even
though the experimental techniques are systematically quite
different, new measurements with improved precision and reduced
systematics are necessary.

The magnetic quenching experiment shown in Fig. 7 is now
being modified with the goal of improving the precision of λ_T to
the 200 ppm level. The field has been increased to 7 kG to
"quench away" the m=0 state by t=150 nsec. The ^{68}Ge positron
source has been replaced with ^{22}Na (which has a lower endpoint
energy) to increase the stopping power of the gas and hence to
increase the Ps formation rate. More γ shielding has been added
and the electronic rejection scheme has been modified to increase
the signal-to-noise ratio in the lifetime spectrum. Preliminary
results are anticipated this fall.

Ps$^-$ Decay

Having first observed the positronium negative ion, Ps$^-$,[26]
Mills has just recently reported[59] a measurement of the Ps$^-$ decay
rate, Γ. Ps$^-$ is formed on a thin carbon film and accelerated by
applying a constant potential, V, to two grids separated by a
distance d. Measurements of the number of ions reaching a the
second grid as a function of d (and hence the proper time since
Ps$^-$ emission) yields Γ. Ps$^-$ decay is detected by its Doppler-
shifted annihilation gammas (see Fig. 1). Measurements were
acquired at accelerating potentials of 1 and 4 kV. The experimental result is $\Gamma = 2.09 \pm 0.09$ nsec^{-1}. This result is in agreement with the two photon Hylleraas-type calculations[60] of Ho
(2.0908 nsec^{-1}) and Bhatia and Drachman (2.0928 nsec^{-1}). These
calculations should be corrected for 3γ annihilation, should
include the 2γ radiative correction, and should allow for bound
state and relativistic effects--all of relative order α.[61]

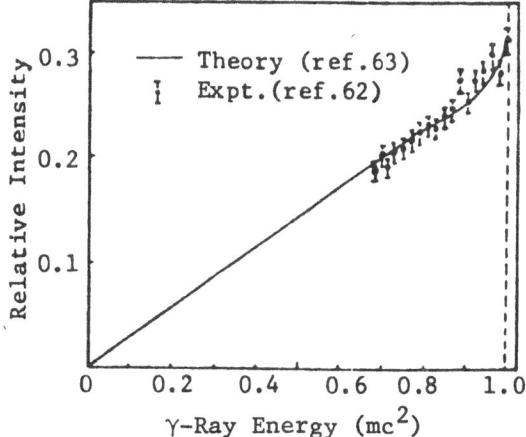

Fig. 8. The measured energy spectrum from triplet Ps decay.
From ref. 62.

3γ Energy Distribution for o-Ps Decay

Chang et al.[62] have measured directly the 3γ energy spectrum
for o-Ps decay, calculated by Ore and Powell in 1949.[63] Their
method uses a β-γ spectrometer, incorporating a thin scintillater
to detect positrons leaving a ^{22}Na source and a HPGe detector for
the annihilation radiation. Their results are shown in Fig. 8.

Positronium Spectroscopy

The energy levels of the n=1 and n=2 states of Ps are shown
in Fig. 9. A milestone in Ps spectroscopy was achieved when Chu
and Mills announced[64] the first optical excitation of Ps. The
1^3S_1-2^3S_1 transition was observed using two-photon Doppler free
excitation. This experiment required the development of high
intensity time "bunched" slow positron beams, efficient thermal Ps
formation in vacuum, and high power tunable lasers near 486 nm.
This is an important measurement in that it should be able to attain
a precision comparable to that of the 1S-2S excitation in hydrogen[65]
and, in so doing, would be a test of QED with accuracy approaching
that of the Ps ground state hyperfine structure (to be discussed).

The Ps-laser interaction region is shown in Fig. 10. In time-
delayed coincidence with a positron pulse a Nd-YAG-dye laser
amplifer delivers an 18 mJ 10 μsec long counter propagating pulse
(at 486 nm) approximately 3 mm in front of a hot Cu(111)+S target.
About 30% of the thermally desorbed Ps from the target, C, intersect
the laser beam L, and of these the photon flux is high enough to
excite the 1S-2S transition with about 30% probability. The 2S Ps

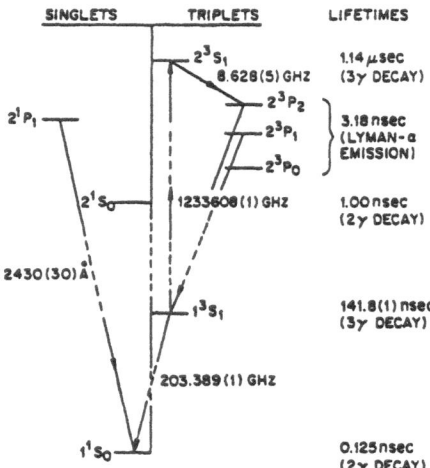

Fig. 9.   Energy levels of the n=1 and n=2 states of positronium.
The quantities with error bars in parentheses are
measured values.   (From ref. 66).

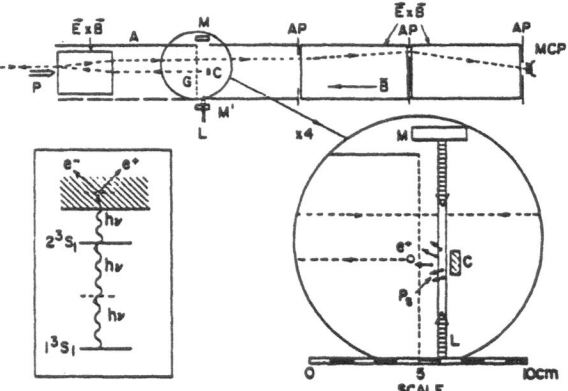

Fig. 10. The positronium-laser beam interaction region (from ref.
64). P is the pulsed positron beam. After the laser beam,
L, excites the 3 photon transition shown in the inset the
ionized Ps fragment follows the dashed line path and is
detected in a multichannel plate array, MCP.

is nearly always ionized, as shown in the insert. The positron
fragment is accelerated by a grid, G, back up the beam into an
E x B region that side steps the positron so that on the return
oscillation it misses the Cu target. This 3 photon ionized Ps is
detected by a multichannel plate, MCP, and can be distinguished
from re-emitted prompt positrons by the 30 nsec delay in the laser
firing. With 10 laser pulses per sec and $\simeq$ 20 positrons per pulse
a signal of about 1.7 cps was observed with 20 to 1 signal-to-noise
and 1.5 GHz linewidth.

The observed resonance frequency is $\Delta\nu$ = 1 233 607 900 $\pm$ 1000
MHz. The theoretical value is[66]

$$\Delta\nu(2^3S_1 - 1^3S_1) = 3/8 \ R_\infty c (1 + k_2\alpha^2 + k_3\alpha^3 \ .....)$$

$$= 1 \ 233 \ 607 \ 197 \pm 1 \ \text{MHz} \pm ?$$

The 1 MHz error is entirely due to the error in R_∞. However, the
uncalculated terms of relative order α^4 could contribute $\pm$ 4 MHz if
k_4 is $\pm$ 1. An improved experiment is now underway at Bell Laboratories that should challenge the theorists to calculate k_4.
Improvements include a new laser with 50 times narrower bandwidth,
a 4-fold improvement in the targetable positron rate, and vastly

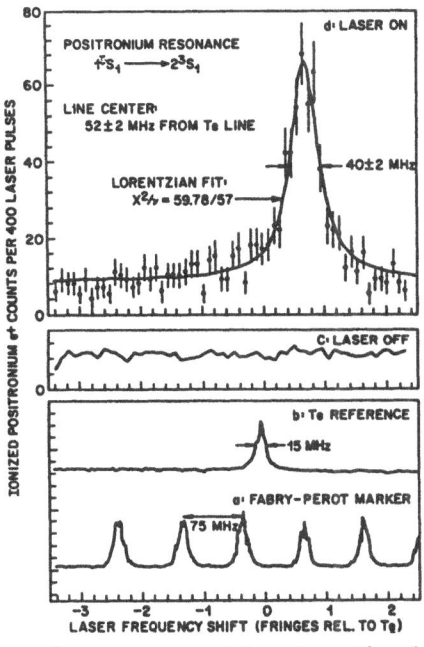

Fig. 11.   The observed resonance line for the 1^3S_1 to 2^3S_1 transi-
           tion (preliminary results from ref. 67). The 40 MHz
           linewidth represents a factor of $\approx$ 50 improvement over
           the earlier experiment reported in ref. 64.

improved laser metrology (the major source of error in the first
experiment). A preliminary result is shown in Fig. 11. The goal
is to measure $\Delta\nu$(1S-2S) with an accuracy below 10 MHz.[67]

The most precise Ps test of QED is the ground state hyperfine
splitting $\Delta\nu$. The theoretical expression for $\Delta\nu$ may be written
as[54,68]

$$\Delta\nu(\text{theory}) = 7/6\ \alpha^2 R_\infty c(1 + h_1\alpha + \ldots.) = 203\ 400\ \text{MHz}$$

where $h_1 = -\dfrac{6}{7\pi}$ (16/9 + ln2) and h_2 includes a term $5/14\ \alpha^2 \ln\alpha^{-1}$.
Uncalculated terms in h_2 of order $\alpha^4 R_\infty c$ corresponding to k_4 in
$\Delta\nu(2^3S_1-1^3S_1)$ could easily contribute $\pm$ 10 MHz to $\Delta\nu$(theory).
Most attention has been directed at revising the previous experimental results. Rich[69] pointed out that inclusion of annihilation
terms in the n=1 Ps Hamiltonian shifts the real part of the Zeeman
eigenvalues by approximately $(\lambda_S/4\pi\Delta\nu)^2$, or 10 ppm. He suggested
that the lineshape calculations be re-investigated. This prompted
Mills[70] to examine the problem in detail. He finds that the
Brandeis[71] and Yale[72] values need to be increased by 2.5 ppm and
21 ppm respectively. The Yale group has just completed an improved
measurement of $\Delta\nu$ in N_2 gas.[73] With an 18 ppm correction to their
previous values, the data are combined to yield the final result.
Therefore the current values for $\Delta\nu$ are:

$\Delta\nu$(Brandeis) = 203 387.5 $\pm$ 1.6 MHz (8 ppm)

$\Delta\nu$(Yale) = 203 389.10 $\pm$ 0.74 MHz (3.6 ppm) .

There are no new results to report on the Ps* fine structure
measurement $\Delta\nu(2^3S_1-2^3P_2)$. We note, for completeness, that the
theoretical value[74]

$$\Delta\nu_{th}(2^3S_1-2^3P_2) = \frac{23}{480}\ \alpha^2 R_\infty c(1 + 3.766\ \alpha + \ldots.) = 8625.14\ \text{MHz}$$

is in agreement with the Brandeis measurement[75] 8631 $\pm$ 6 MHz. An
improved experiment is underway at Brandeis and their search[30] for
a surface with enhanced Ps* formation discussed earlier in this
paper will hopefully allow them to attain their goal of $\pm$ 1 MHz in
$\Delta\nu(2^3S_1-2^3P_2)$.

In conclusion, we note that there has been steady progress in
measuring and calculating the fundamental properties of Ps. Two-
photon optical excitation of Ps has finally been realized. The
radiative correction to λ_S have finally been checked. The decay
rate of Ps$^-$ has been measured for the first time and two new
independent calculations of Γ of sufficient accuracy are in agreement with the measurement. The ground state hyperfine structure
lineshape question appears to have been resolved with one of the

values changing substantially. On the other hand, the triplet decay
rate measurements are still not in satisfactory agreement with
theory. Finally, the most significant theoretical deficiency at
this time is the lack of order $\alpha^4 R_\infty c$ computations to the Ps
spectroscopy. The spectroscopic measurements of $\Delta\nu, \Delta\nu(1^3S_1-2^3S_1)$,
and $\Delta\nu(2^3S_1-2^3P_2)$ have either attained or soon will obtain
accuracies of 1-10 MHz- just exactly the uncertainty attributable
to the uncalculated terms.

ACKNOWLEDGEMENTS

We thank R. Drachman, A. Köymen, M. Leventhal, K. Lynn,
A. Mills, A. Rich, D. Schoepf, S. Sharma, and P. Zitzewitz for
helpful discussions. Positronium work at The University of Texas at
Arlington and at The University of Michigan is funded by the
Division of Atomic and Molecular Physics of the National Science
Foundation. The support of the Robert A. Welch Foundation for the
UTA program is also gratefully acknowledged. One of us (D.W.G.) is
the recipient of an Alfred P. Sloan Fellowship.

REFERENCES

1. D. W. Gidley, Can. J. Phys. 60, 543 (1982).
2. D. W. Gidley, A. Rich, and P. W. Zitzewitz, in Positron
    Annihilation, P. G. Coleman, S. C. Sharma and L. M. Diana,
    eds., p. 11, North-Holland (1982).
3. A. Rich, Rev. Mod. Phys. 53, 127 (1981).
4. S. Berko and H. N. Pendleton, Ann. Rev. Nucl. Part. Sci. 30,
    543 (1980).
5. P. G. Coleman, T. C. Griffith, G. R. Heyland and T. L. Killeen,
    J. Phys. B. 8, L185 (1975).
6. S. C. Sharma, J. D. McNutt, A. Eftekhari, and Y. J. Attaiiyan,
    Can. J. Phys. 60, 610 (1982).
7. T. C. Griffith, M. Charlton, G. Clark, G. R. Heyland, G. L.
    Wright, in Positron Annihilation, P. G. Coleman, S. C. Sharma,
    L. M. Diana, eds., p. 61, North-Holland (1982).
8. O. E. Mogensen, ibid, p. 763 and F. M. Jacobsen, N. Gee, and
    G. R. Freeman, ibid, p. 92.
9. M. Charlton, T. C. Griffith, G. R. Heyland, K. S. Lines, G. L.
    Wright, J. Phys. B. 13, L757 (1980).
10. D. R. Cook, P. G. Coleman, L. M. Diana, S. C. Sharma, in
    Positron Annihilation, P. G. Coleman, S. C. Sharma, L. M.
    Diana, eds. p. 86, North Holland (1982).
11. J. Humberston, Can. J. Phys. 60, 591 (1982).
12. A. S. Ghosh, N. C. Sil, P. Mandal, Phys. Reports 87, 313 (1982).
13. R. Paulin and G. Ambrosino, J. Physique 29, 263 (1968).
14. S. M. Curry and A. L. Schawlow, Phys. Lett. A 37, 5 (1971).

15. K. F. Canter, P. G. Coleman, T. C. Griffith, G. R. Heyland, J. Phys. B 5, L167 (1972).

16. G. W. Ford, L. M. Sanders, T. A. Witten, Phys. Rev. Lett. 36, 1269 (1976).

17. D. W. Gidley, K. A. Marko, A. Rich, Phys. Rev. Lett. 37, 729 (1976).

18. S. Y. Chuang and S. J. Tao, Can. J. Phys. 51, 820 (1973).

19. S. M. Kim and W. J. L. Buyers, J. Phys. C. 11, 101 (1978).

20. K. F. Canter, A. P. Mills, Jr., S. Berko, Phys. Rev. Lett. 33, 7 (1974).

21. A. P. Mills, Jr., Phys. Rev. Lett. 41, 1828 (1978).

22. A. P. Mills, Jr. and L. Pfeiffer, Phys. Rev. Lett. 43, 1961 (1979).

23. R. M. Nieminen and J. Oliva, Phys. Rev. B 22, 2226 (1980).

24. K. G. Lynn and D. O. Welch, Phys. Rev. B 22, 99 (1980).

25. S. Chu, A. P. Mills, Jr., C. A. Murray, Phys. Rev. B 23, 2060 (1981), and R. M. Nieminen and M. J. Puska, Phys. Rev. Lett. 50, 281 (1983).

26. A. P. Mills, Jr., Phys. Rev. Lett. 46, 717 (1981).

27. K. G. Lynn, private communication.

28. A. P. Mills, Jr., private communication.

29. K. F. Canter, A. P. Mills, and S. Berko, Phys. Rev. Lett. 34, 177 (1975).

30. D. C. Schoepf, S. Berko, K. F. Canter, and A. H. Weiss, in Positron Annihilation, P. G. Coleman, S. C. Sharma, and L. M. Diana, eds. p. 165, North-Holland (1982).

31. D. W. Gidley, A. R. Köymen, and T. W. Capehart, Phys. Rev. Lett. 49, 1779 (1982).

32. J. Van House and P. W. Zitzewitz, private communication.

33. M. Leventhal, C. J. MacCallum, and P. D. Stang, Astrophys. J. 225, L11 (1978), and references therein.

34. G. R. Riegler, J. C. Ling, W. A. Mahoney, W. A. Wheaton, J. B. Willett, A. S. Jacobson, and T. A. Prince, Astrophys. J. 248, L13 (1981).

35. M. Leventhal, C. J. MacCallum, A. F. Huters, and P. D. Stang, Astrophys. J. 260, L1 (1982).

36. W. S. Paciesas, T. L. Cline, B. J. Teegarden, J. Tueller, P. Durouchoux, and J. M. Hameury, Astrophys. J. 260, L7 (1982).

37. R. W. Bussard, R. Ramaty, and R. J. Drachman, Astrophys. J. 228, 928 (1979), and R. J. Drachman, in Positron Annihilation, P. G. Coleman, S. C. Sharma, and L. M. Diana, eds., p. 37, North Holland (1982).

38. K. G. Lynn, in Proceedings of International School of Physics "Enrico Fermi", 83rd session, Varenna, Italy, June 1981, W. Brandt, A. Dupasquier, eds. (Academic Press, N.Y., 1983).

39. K. G. Lynn, Phys. Rev. Lett. 43, 391 (1979).

40. K. G. Lynn, Phys. Rev. Lett. 44, 1330 (1980).

41. H. H. Jorch, K. G. Lynn, I. K. MacKenzie, Phys. Rev. Lett. 47, 362 (1981).

42. D. W. Gidley, A. Rich, J. C. Van House, and P. W. Zitzewitz, Nature 297, 639 (1982).
43. R. A. Hegstrom, Nature 297, 643 (1982).
44. S. C. Sharma, A. Eftekhari, J. D. McNutt, Phys. Rev. Lett. 48, 953 (1982).
45. A. P. Mills and S. Berko, Phys. Rev. Lett 18, 420 (1967).
46. K. Marko and A. Rich, Phys. Rev. Lett. 33, 980 (1974).
47. G. P. Lepage and P. B. Mackenzie, private communication.
48. G. S. Adkins and F. R. Brown, Princeton University preprint (1983) and references therein.
49. I. Harris and L. Brown, Phys. Rev. 105, 1656 (1957).
50. V. K. Cung, A. Devoto, T. Fulton, and W. W. Repko, Phys. Rev. A 19, 1886 (1979).
51. J. R. Freeling, Ph.D. Thesis, University of Michigan (1979).
52. Y. Tomozawa, Annals of Physics 128, 463 (1980).
53. D. W. Gidley, A. Rich, E. Sweetman, and D. West, Phys. Rev. Lett. 49, 525 (1982).
54. W. E. Casewell and G. P. Lepage, Phys. Rev. A 20, 36 (1979).
55. D. W. Gidley, A. Rich, P. W. Zitzewitz, and D. A. L. Paul, Phys. Rev. Lett. 40, 737 (1978).
56. T. C. Griffith, G. R. Heyland, K. S. Lines, and T. R. Twomey, J. Phys. B 11, 743 (1978).
57. D. W. Gidley and P. W. Zitzewitz, Phys. Lett. A 69, 97 (1978).
58. M. A. Stroscio, Phys. Rev. Lett. 48, 571 (1982).
59. A. P. Mills, Jr., Phys. Rev. Lett. 50, 671 (1983).
60. Y. K. Ho, J. Phys. B 16, 1503 (1983) and refs. therein.
61. A. K. Bhatia and R. J. Drachman, Phys. Rev. A, to be published. The value quoted in this paper, $\Gamma = 2.0861$ nsec^{-1}, includes 2γ radiative and 3γ annihilation corrections.
62. Chang T-B, Li Y-Q, Liu N-Q, Wang Y-Y, Tang X-W, in Positron Annihilation, P. G. Coleman, S. C. Sharma, L. M. Diana, eds. p. 32, North Holland (1982).
63. A. Ore and J. L. Powell, Phys. Rev. 75, 1696 (1949).
64. S. Chu and A. P. Mills, Phys. Rev. Lett. 48, 1333 (1982).
65. C. Wieman and T. W. Hänsch, Phys. Rev. A 22, 192 (1980).
66. A. P. Mills and S. Chu, Proceedings of the 8th International Conf. on Atomic Physics, Göteborg, Sweden, Aug. 1982, and references therein.
67. S. Chu, J. Hall and A. P. Mills, Jr., private communication.
68. R. Karplus and A. Klein, Phys. Rev. 87, 848 (1952).
69. A. Rich, Phys. Rev. A 23, 2747 (1981).
70. A. P. Mills, Jr., Phys. Rev. A 27, 262 (1983).
71. A. P. Mills and G. H. Bearman, Phys. Rev. Lett. 34, 246 (1975).
72. P. O. Egan, V. W. Hughes, and M. H. Yam, Phys. Rev. A 15, 251 (1977).
73. M. W. Ritter, P. O. Egan, V. W. Hughes, and K. A. Woodle, Yale University preprint (1983).
74. T. Fulton and P. C. Martin, Phys. Rev. 95, 811 (1954).
75. A. P. Mills, S. Berko, and K. F. Canter, Phys. Rev. Lett. 34, 1541 (1975).

76. T. Muta and T. Niuya, Progress of Theoretical Physics 68,
     1735 (1982) and A. Billoire, R. Lacaze, A. Morel and H.
     Navelet, Phys. Letters 78B, 140 (1978).
77. A different approach to this calculation is presented in
     W. Celmaster and D. Sivers, Phys. Rev. D23, 227 (1981).
     Their result is 0.024% higher.

POSITRONIUM FORMATION IN GASES AND LIQUIDS

Finn M. Jacobsen

Chemistry Department
Riso National Laboratory
DK-4000 Roskilde
Denmark

Abstract

Generally, most discussions of the yields of positronium, Ps, in gases and liquids have been constrained to the framework of either the Ore or spur model of Ps formation. In this paper, we shall demonstrate that by a slight extension of the frames of these models it is possible to formulate a single model of Ps formation in gases and liquids. Furthermore, we shall give a semi-quantitative discussion of the thermalization properties of positrons in gases. These results will be used in a discussion of the density behaviour of Ps yields in gases.

1. INTRODUCTION

For almost a decade the subject of a number of discussions has been: which of the two models of positronium, Ps, formation, the Ore[1] or spur model[2], gives the best description of Ps formation processes in liquids[2-10]. In gases, on the other hand, most discussions of Ps yield results have been carried out within the framework of the Ore model[11-17]. It has been acknowledged for some time, however, that the Ore model, at least in its simplest version, is capable of explaining the strong density dependence of the Ps yield as observed in many gases[15-17]. It has been pointed out several times that it is likely that spur processes also have an impact on the Ps yield in high density gases (> 10 amagat)[4,18,19] and that this perhaps could be part of the explanation of the density behaviour of the Ps yield in most gases.

Although some scientists favour the Ore model combined with hot Ps chemistry as the best picture of the nature of Ps formation

processes in liquids, it is probably fair to say that the current
state-of-the-art concerning Ps formation is that the majority of
scientists agree that the Ore model is a sound picture of the Ps
formation processes in low density gases, while very few doubt that
spur processes have a significant impact on the Ps yield in high
density liquids.

A close inspection of how well we understand the Ps formation
processes in pure gases and liquids on the one hand is very pessi-
mistic, but on the other is rather optimistic. This latter view
refers to the fact that we have still a lot to learn about the
physical properties of positrons and electrons in matter. Concerning Ps formation in gases, we are able to explain the Ps yield in
the three lightest rare gases He, Ne, and Ar fairly well[20], while
our understanding of the Ps yields in Kr and Xe, not to mention
molecular gases, is very uncertain[11,12,20]. In liquids we encounter
a similar level of understanding of the Ps yields[21,22]. Although,
the spur model has proved to be very successful in explaining Ps
yields in liquid solutions[4,23-25] and is able to account for some
gross differences of the Ps yield between various groups of liquids,
we are still far away from understanding the Ps yield in pure
liquids in detail.

In this paper we shall discuss Ps formation processes in gases
and liquids without referring to a specific model of Ps formation,
but rather to general properties such as the slowing down of light
particles, hot Ps, spurs, etc.

In order to make the paper reasonably self-consistent, the Ore
and spur model will be described in section 2.1 and 2.2, respectively.
Section 3.1 contains a general discussion of Ps formation processes
in gases and liquids while in section 3.2 we shall give a more
quantitative discussion of the Ps formation processes in some
selected cases. Section 4 contains a summary and conclusions.

2. MODELS OF Ps FORMATION PROCESSES

The purpose of this section is to state the essential features
of the Ore and spur models of Ps formation, rather than repeat
discussions of Ps formation in terms of these models previously
given by a number of authors.

2.1 THE ORE MODEL OF Ps FORMATION

In the Ore model Ps formation is treated as one of the inelastic collision channels in a positron molecule (atom) collision.
As the binding energy of Ps is 6.8 eV, Ps can be formed as a result
of a positron-molecule collision only for positron energies in
excess of I-6.8 eV, I being the ionization potential. For positron
energies above I, the ionization process is often assumed to be much

more likely than that of Ps formation. The argument has been that
the ionization process contains three particles in the final state
while that of Ps formation only two. In the noble gases it is
unnecessary to apply the above argument to place an upper limit on
the positron energy above which Ps formation can be neglected. If
Ps is formed with an excess energy above 6.8 eV in the noble gases
it is assumed that part of the Ps formed will probably break up
before being moderated down to energies where dissociation is no
longer possible. This is because the energy moderation processes
of Ps are very inefficient in the noble gases; the fractional
average energy loss per collision is $2m_{Ps}/M$, where M is the mass
of the atom. Thus, in terms of the Ore model we need only consider
Ps formation for positron energies, E_p, in the interval of
$I-6.8 < E_p < I$. In this energy interval Ps formation competes with
other inelastic collision processes (elastic collisions can be
neglected because of very low moderation power) and we can formally
express the Ps formation probability, P, vs the positron energy as:

$$P(E) = \begin{cases} \dfrac{\sigma_{Ps}(E)}{\sigma_{Ps}(E) + \Sigma\sigma_{in}(E)} & ; \quad I-6.8 < E < I \text{ eV} \\ \\ 0 & ; \quad \text{otherwise} \end{cases} \qquad (1)$$

σ_{Ps} being the cross-section for Ps formation and $\Sigma\sigma_{in}$ that for the
inelastic collision channels. Essentially Eq. 1 expresses what is
known as the Ore model of Ps formation. In most cases, however, we
have insufficient information on the cross-section terms to be able
to make use of Eq. 1. On the other hand, we can take advantage of
the fact that in most experimental set-ups the initial distribution
of positron energies is such that we can assume that the probability
that a positron will slow down to an energy E in the interval I-kT
can be considered as independent of E. Under these circumstances
Eq. 1 can provide us with a lower, F_1 and an upper, F_u, estimate of
the Ps yield. The lower estimate, F_1, is obtained by assuming
$\Sigma\sigma_{in} \gg \sigma_{Ps}$

$$F_1 = \begin{cases} \dfrac{E^* - I + 6.8}{I} & ; \quad E^* > I-6.8 \text{ eV} \\ \\ 0 & ; \quad E^* < I-6.8 \text{ eV} \end{cases} \qquad (2)$$

E^* being the lowest excitation energy. The upper estimate is
obtained by reversing the condition on the cross-sections:

$$F_u = 6.8/I \text{ eV} \qquad (3)$$

Most Ps yields in low density gases are found to be roughly within
the limits given by Eqs. 2 and 3; important exceptions are Kr and
Xe[12]. For the three lightest gases He, Ne, and Ar, Schrader et

al[20] have used Eq. 1 together with the best available estimates of the cross-section terms and obtained excellent agreement with experimental values of the Ps yields.

2.2 THE SPUR MODEL OF Ps FORMATION

The spur model of Ps formation has been applied mainly to account for Ps yields measured in liquids. In the spur model Ps is assumed to be formed by a reaction between a thermalized positron and one of the excess electrons in the positron spur formed at the end of the positron track. At its formation the positron spur is typically assumed to consist of a number of positive ions (2-4), a corresponding number of excess electrons, the positron, and some radical pairs. Some typical reactions that can take place in the positron spur in nonpolar liquids are shown below:

$$e^{+**} + M_n \rightarrow [e^{+*}, mM^+, me^{-*}] \tag{4a}$$

$$[e^{\pm*} + M_n \rightarrow e^{\pm} + M_n^*] \tag{4b}$$

$$[e^- + M^+ \rightarrow M^*] \tag{4c}$$

$$[M^* + M_n \rightarrow M + M_n^*] \tag{4d}$$

$$[e^+, e^-, M^+] \rightarrow e^+, e^-, M^+ \tag{4e}$$

$$[e^+ + e^- \rightarrow Ps] \tag{4f}$$

$$[e^+ + M^* \rightarrow Ps + M^+] \tag{4g}$$

$$[Ps + M^+ \rightarrow e^+ + M] \tag{4h}$$

$$[e^- + Ps \rightarrow Ps^-] \tag{4i}$$

$$[e^+ + M_n \rightarrow q\gamma + M_{n-1}M^+] \tag{4j}$$

the [ ] indicate that the processes take place inside the spur. In nonpolar liquids the most important processes as far as Ps formation is concerned are 4c, e, f, j, where c symbolizes recombination of excess electrons and positive ions, e: diffusion out of the spur of e^+, e^-, M^+ to become free ions, f: Ps formation, and j: decay of the "free" positron. In a first approximation the Ps yield depends on the relative strength of 4f compared to that of 4c, e, and j. Of secondary importance are 4d, g, h, and i. In polar liquids Ps formation in presolvated states of e^+ and/or e^- and by encounter pair reactions of e^- and e^+ have to be considered, as the mobility of e^- and e^+ in the solvated state is too low in many cases to allow for Ps formation within the lifetime of the "free" positron.

Although the Ps yields in various liquids correlate to spur properties such as the thermalization distance of secondary electrons, the spur lifetime and to properties such as the work function and mobility of excess electrons in liquids[21],[23] the experimental support of the spur model of Ps formation has come about through studies of the Ps yield in solutions of liquids. By adding electron accepting molecules (A) to a liquid we can put the Ps formation process 4f in competition with: $e^- + A \rightarrow A^-$. Let A denote a strong electron acceptor and a a weak one. By adding a and A to a liquid we can introduce the following processes:

$$[e^- + A \rightarrow A^- \quad] \qquad\qquad (5a)$$

$$[e^+ + A^- \rightarrow [e^+,A^-]] \qquad\qquad (5b)$$

$$[e^- + a \rightarrow a^- \quad] \qquad\qquad (5c)$$

$$[e^+ + a^- \rightarrow Ps + a] \qquad\qquad (5d)$$

Processes 5a and b represent Ps inhibition while that of 5c the anti-recombination effect. The competition between 5a and c gives rise to the anti-inhibition effect. One of the most studied solvents with respect to the processes (5) is hexane[22]. Another effect that illustrates the influence of spur processes on the Ps yield is that of dehalogenation time on the Ps yield[25].

3 Ps FORMATION PROCESSES IN GASES AND LIQUIDS

Ps yield investigations in gases have often been carried out in a density region from 1 amagat to above the critical density[14],[17] (in the super critical region). Recently, the corresponding experiments in the liquid state have started to cover the temperature range from the melting point to above the critical point[22],[26]. Here we shall discuss Ps yield in gases and liquids without making any sharp distinction between the two phases. This is of course, not to say that we do not expect marked differences in the efficiency of various physical processes that determine the measured Ps yield in the extreme density limits of the two phases, but just that we do not expect a sharp change of the Ps yield at the critical point.

In this section we shall discuss the Ps formation processes in the density region from diluted gases to high-density liquids. In section 3.1 we shall discuss Ps formation in general terms while in section 3.2 we shall give a more quantitative account of the Ps yield in some selected cases.

3.1 A GENERAL DISCUSSION OF Ps FORMATION IN GASES AND LIQUIDS

It is of interest to note what small changes are needed to the
two models of Ps formation, the Ore and spur models, in order to
transform them into one. Roughly speaking, if, in the Ore model,
we allow for moderation of "hot" Ps ($E_{kin} > 6.8$ eV) in a very
general sense, and in the spur model open up the possibility of Ps
formation in direct collisions for positron energies below the
ionization threshold of the medium, the two models become very much
alike. The advantage of making these extensions of the two models
is obvious as we now are forced to discuss Ps yield results in
terms of more basic properties like the moderation of hot Ps, e^+,
and e^-, the efficiency of Ps production in a direct collision below
the ionization threshold, secondary reactions of Ps, etc. as functions of density and temperature. In a perhaps somewhat restricted
sense, the moderation of hot Ps combined with the Ore model has been
suggested by several authors in order to account for the density
dependence of the Ps yields in gases[11],[17]. In the spur model, on
the other hand, most authors in favour of this model have not given
much attention to the possibility of Ps production by an Ore type
mechanism. However, to the author's knowledge there exist no
fundamental arguments against Ps formation in a direct collision
between a positron and a molecule below the ionization threshold
even in high-density liquids. One may argue that studies of the
measured Ps yield in solutions of liquids support the conclusion
that only spur processes are important as far as the Ps yield is
concerned. For example, by adding a strong electron acceptor such
as CCl_4 to hexane it is possible to inhibit Ps formation completely[22].
The explanation offered is that the Ps formation processes compete
with the reaction: $e^- + A \rightarrow A^-$. Clearly, however, the effect of
adding any electron acceptor to a liquid or gas is twofold: One,
to decrease the probability of Ps formation by a reaction between
a positron and a spur electron, another to increase the likelihood
of oxidation of Ps by positive ions. Of course the latter process
does not depend directly on the Ps formation mechanism. Oxidation
of Ps by positive ions has been discussed as one mechanism that
can explain why the Ps yield in pure liquids as determined in lifetime and angular correlation experiments are different in many
cases[22]. Furthermore, very recently, high-resolution lifetime
experiments in pure liquids have shown that a typical lifetime
spectrum is not described very accurately by just three exponen-
tially decaying lifetime components[27].

Although we do not intend to discuss the results in detail,
we show in Figure 1 the yield of the escaped o-Ps from the surface
of crystalline ice vs the energy of the incoming positron[28]. Note
the similarity of the results displayed in Figure 1 at low positron
energies with the Ps formation cross-section vs positron energy in
very low-density gases[29]. Although, in the interpretation of the

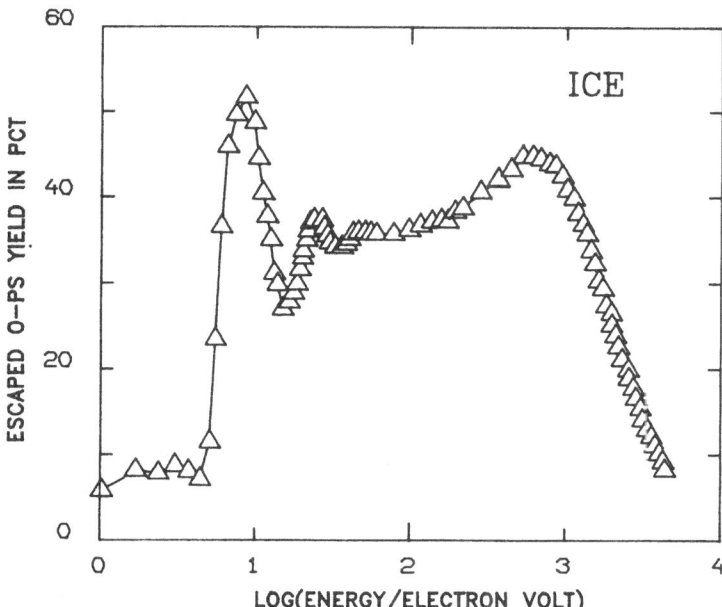

Fig. 1:   The escaped yield of o-Ps from a crystalline ice surface
vs the energy of the incoming positron; T=150K. (The
curve has been read off a figure from ref. 28).

results shown in Figure 1, the surface plays a significant role,
in no way do the results in Figure 1 indicate a marked decrease
of the efficiency of Ps production by an Ore type mechanism in very
dense matter. Qualitatively, the surface can influence the escaped
o-Ps yield in two ways: 1) At positron energies where spur formation begins to take place the spurs are cut by the surface, and
2) for Ps formed close to the positive ion and close to the surface,
the surface can compete with the positive ion for a reaction with
Ps.

The strong density dependence of the Ps yields in gases, and
the new results for Ps formation cross-sections in very low density
gases[29], together with the discussion and remarks given above lead
us to suggest that in a discussion of Ps yields in gases and liquids
the following mechanisms for production of thermalized Ps should be
taken into account:

$$e^{+*} + M \rightarrow Ps + M^+ \tag{6a}$$

$$e^{+**} + pM \rightarrow \begin{cases} Ps^* + pM + M^+ \rightarrow Ps + pM^* + M^+ & \text{(6b)} \\[2ex] (e^+, e^-)^* + pM + M^+ \rightarrow Ps + pM^* + M^+ & \text{(6c)} \\[2ex] m(e^- + M^+) + e^+ \rightarrow Ps + (q-1)e^- + qM^+ & \text{(6d)} \end{cases}$$

pM represents the gas or liquid molecules. 6a symbolizes Ps formation in a direct collision for E_{kin} I, while 6b, c, and d show, respectively, moderation of hot Ps, correlated motion of an electron positron pair during slowing down, and Ps formation by a spur reaction. It must be emphasized that in (6) we have not shown any of the many Ps^*, e^+, and e^- reactions that may act as competitive reactions to those listed in (6). The three last reactions in (6) must not be taken as completely separated pathways for production of thermalized Ps. For example, one obvious question we can ask is: how much Ps character should a $(e^+, e^-)^*$ pair take to belong to 6b rather than to 6c or d? The main reason for dividing the process $e^{+**} + pM$ into three sub-reactions is to be found in the effect of density and competitive reactions on the efficiency of $e^{+**} + pM \rightarrow Ps$.

The slowing-down properties of e^+, e^-, and Ps play an essential role for the efficiency of 6a-d. In the following we shall try to obtain a semi-quantitative estimate of the thermalization time and distance of e^+, e^-, and Ps in gases. We shall approach the problem in a way similar to that in ref. 30.

Let E be the average energy of the particle, λ the mean free path, f the average fractional energy loss per collision, and E_{th} the thermal energy: then we can express the rate of change of the average energy as:

$$\frac{d}{dt}(E) = -(2/m)^{1/2}(f/\lambda)E^{1/2}\{E - E_{th}\} \tag{7}$$

m being the mass of the particle. In general, both λ and f depend on E; however, in the present calculation we shall be satisfied by using some appropriate constant average value of λ/f. By defining the thermalization time, τ, as $E(\tau) = 1.1E_{th}$ we obtain from Eq. 7:

$$\tau \sim (\lambda/f)(m/2)^{1/2}\{3.7(E_{th})^{-1/2} - 2(E_o)^{-1/2}\} \tag{8}$$

E_o being the energy of the particle at t=0. From Eq. 8 we obtain the density normalized thermalization time as:

$$N\tau = n^{1/2}/(\sigma f) \; \{2.1T^{-1/2} - 1.3 \; 10^{-2}(E_o)^{-1/2}\} \qquad (9)$$

$$\text{amagatnsec}$$

where n is defined as the particle mass in the unit of the mass of the electron and σ is the collision cross-section. The units of σ, T, and E are, respectively, $\text{\AA}^2$, K, and eV.

The thermalization distance, R, can be expressed as:

$$R = \lambda(n_c)^{1/2} \qquad (10)$$

n_c being the number of collisions needed to moderate the particle down to $1.1E_{th}$. n_c can be calculated as:

$$n_c = 1/\lambda (2/m)^{1/2} \int_0^\tau E^{1/2}(t)\,dt \qquad (11)$$

By using $\tau = 3.7 \; \lambda/f(m/2E_{th})^{1/2}$ (from Eq. 8) together with Eqs. 7, 10, 11 we can in a reasonable approximation, express the density-normalized thermalization distance as:

$$RN = 2.6 \; 10^4 (N\tau/\sigma)^{1/2}(T/n)^{1/4}(\ell n(10E_o/E_{th}))^{1/2} \qquad (12)$$

$$\text{\AA amagat}$$

with the same units as those used in Eq. 9. The advantage of the method described above for estimating the thermalization distance is that all information needed about the efficiency of the inelastic scattering processes enter through the density normalized thermalization time which in principle can be determined from experiments. The main problem in using Eq. 12 is to estimate an appropriate value for the total scattering cross-section. In order to facilitate a test of the present calculation we can calculate the density normalized thermalization time in the noble gases where f is known. The results are listed in Table I together with the experimental values:

Table 1: Density-normalized thermalization times of positrons in the noble gases (amagatnsec).

Gas	$< \sigma(\text{A}) >^{31}$	$(N\tau)_{cal}$	$(N\tau)^{12}_{exp}$
He	0.16±0.04	2500± 500	1700± 50
Ne	0.3 ±0.2	7800±3000	1700±200
Ar	6 ±2	650± 100	362± 5
Kr	8 ±4	1500± 900	325± 6
Xe	26 ±6	560± 100	178± 3

Generally, the calculated values of $N\tau$ exceed those determined from experiments by a factor of 1-3. The main source of error in the calculated $N\tau$ values is probably the determination of $<\sigma>$. In the estimate of $<\sigma>$ we should pay relatively more attention to the values of σ at low energies. Unfortunately, there exist today few experimental values of σ below 1 eV. Another source of error that can explain part of the difference between $(N\tau)_{cal}$ and $(N\tau)_{exp}$ is contamination of the gases by polyatomic molecules. Typically $(N\tau)_{exp}$ values in the noble gases can be sensitive to impurities at a level of a few p.p.m. With the above remarks in mind we conclude that the level of agreement between calculated and experimental values of $N\tau$ is as expected.

In the calculation of RN the selected value of E_o is not critical. By assuming T=300 K we obtain from Eq. 12:

$$RN = 2 \ 10^5 (N\tau/\sigma)^{1/2} \ \mathring{A}amagat \tag{13}$$

We show in Table II calculated values of RN for a number of gases together with the values of N^* at which $(1-\exp(-r_c/R)) = 0.1$; r_c is the distance at which the potential energy of a charge pair is equal to kT and $\exp(-r_c/R)$ is the probability that a charge pair of opposite sign separated by a distance R escape recombination[32].

Table II: Density normalized thermalization distance of positrons in gases ($\mathring{A}$amagat)

Gas	$(N\tau)_{exp}$[12]	$<\sigma>$[31,33]	$(RN)_{cal}$	N^*
Xe	178	26±6	$(5\pm0.3)10^5$	100
N_2	16	4±1	$(4\pm0.5)10^5$	77±10
H_2	3.3	2±1	$(2.6\pm0.5)10^5$	50±10
CH_4	0.25	5+5	$(4.5-1.4)10^4$	9-3
CO_2	0.09	10+5	$(1.9-0.3)10^4$	3.7-.7
iC_4H_{10}	0.05	10[a]	$1.4 \ 10^4$	2.7

a)   assumed value

In order to estimate the contribution to the Ps yield from processes of the type 6c and d we should carry out similar calculations for thermalization of electrons. However, from experiments we know that the corresponding values of $N\tau$ and RN for electrons are comparable to or smaller than those shown for positrons in Table II[30]. It should be emphasized that in the present calculation of RN we have disregarded the effect of the Coulomb fields. Furthermore the cross-sections that enter into the calculation are total and not momentum transfer cross-sections.

The values of N^* in Table II indicate that we can expect to observe density effects on the Ps yield at densities from a couple of amagat in molecular gases. Although we have disregarded the positive ion, the value of N^* are probably mostly associated with the contributions to the Ps yield from randomly (non-correlated) distributed thermalized electron-positron pairs (spur effect; 6d). If, initially, the electron-positron pairs are strongly correlated (6b and c) we expect the density effect on the Ps yield to set in at lower density than N^*.

3.2 DISCUSSION OF THE Ps YIELD IN SOME SELECTED GASES

One way to compare with the experimental values of the Ps yields in gases is to compare the density at which the Ps yields are equal to $F_{low} + 0.1(1 - F_{low})$ with that of N^*; F_{low} being the Ps yield at such low density that no density effect is expected. The comparison is displayed in Table III:

Table III: Comparison of the density N^*_{exp} where $F_{exp} = F_{low} + 0.1(1 - F_{low})$ with N^* from Table II.

Gas	N_2	H_2	CH_4	CO_2
N^{*a}	77±10	50±10	9–3	3.7–0.7
N^*_{exp} a	40^{15}	22^{17}	8^{34}	3.5^{34}

a) N^* in amagat units; T=300 K

The calculated values of N^* agree surprisingly well with the experimental ones for CH_4 and CO_2. However, it should be mentioned that the experimental values used for F_{low} for these two gases may not have been determined at a low enough density. Based on the nature of the calculation of N^* it may turn out to be somewhat speculative to discuss the differences between the N^* and N^*_{exp} values for N_2 and H_2. Recently, however, experimental results on the 3γ yield vs positron energy[29] show that the correlation of e^+ and e^- in an ionization process above the ionization threshold can be very strong indeed. So perhaps the differences between N_{exp} and N^* contain information about the Ps formation processes 6b ,c.

By comparing Eqs. 9 and 12 we observe that RN is not expected to depend very much on the temperature. However, the value of r_c and hence N^* does. By lowering the temperature to 77 K the values of N^* are decreased by a factor of 3.9 compared with those shown in Table II and III. By changing the temperature from 293 K to

77 K the observed behaviour of the Ps yield in H_2[17] is in qualitative agreement with this prediction.

In the above discussion we have not taken the finite lifetime
of the positron into account. However, the efficiency of the Ps
formation channels 6c and d can be sensitive to the decay probability of e^+. The Ps formation time, t_{Ps}, vs the distance, R, between
the electron and positron is given roughly by[7,21]:

$$t_{Ps} = 4/3\epsilon_o\epsilon_r R^3/(e\mu) \tag{14}$$

μ being the sum of the mobilities of the electron and positron. If
t_{Ps} associated with 6c and d becomes comparable to the decay rate
of the positron these Ps formation processes beomme less efficient.
This may partly explain why the Ps yields decrease at high density
in $N_2(297K)$[15], $SF_6(297K)$[12], and $H_2(77K)$[17].

4.   SUMMARY AND CONCLUSIONS

We have discussed a number of experimental results on the Ps
yield in matter that suggest that it may be difficult to interpret
Ps yields in terms of only one of the two models (the Ore and spur
model) of Ps formation. We have investigated in some detail the
possibility of Ps formation in gases from processes other than those
of a direct collision between a positron and a molecule. The conclusion between various types of e^+, e^- pairs which do not initially
have a Ps character. Furthermore, we have questioned the assumption
that Ps cannot be formed in an Ore type of mechanism in very dense
matter. Because of lack of space it has not been possible to discuss
in detail secondary Ps reactions, some of which cause errors in the
determination of the Ps yields from experiments. We shall discuss
this aspect in detail in a forthcoming paper.

ACKNOWLEDGEMENT

The author is very grateful to O.E. Mogensen and M. Eldrup
for many stimulating discussions.

REFERENCES

1. A. Ore, Univ. i Bergen, Årbok, Naturvidenskabelig
   Rekke 9:15pp (1950)
2. O. E. Mogensen, J. Chem. Phys., 60:988 (1974)
3. O. E. Mogensen, Appl. Phys., 6:315 (1975)
4. O. E. Mogensen, "Positron Annihilation", Edited by
   P. G. Coleman, S.C. Sharma and L.M. Diana, North-Holland
   Publ. Co., Amsterdam, (1982) 763.
5. G. Wikander, Chem. Phys., 39:309 (1979).
6. B. Levay and O.E. Mogenson, J. Phys. Chem., 81:373 (1977).
7. F.M. Jacobsen, Ph.D. Thesis, Riso-R-433, (1981) pp 103.

8. H.J. Ache, "Positronium and Muoniom Chemistry", Edited by H.J. Ache, Advances in Chemistry series 175, Washington, (1979), 1

9. C.D. Jonah, J.C. Abbe, G. Duplatre, and A. Haessler, Chem. Phys. 58:1 (1981).

10. A.F. Para and E. Lazzarini, J. Inorg. Nucl. Chem., 40:1473 (1978).

11. T.C. Griffith and G.R. Heyland, Phys. Reports, 39:171 (1978)

12. G.R. Heyland, M. Charlton, T.C. Griffith and G.L. Wright, Can.J.Phys. 60:503 (1982).

13. P.G. Coleman, T.C. Griffith, G.R. Heyland and T.L. Killeen J. Phys. B. 8:1734 (1975).

14. S.C. Sharma and J.D.McNutt, Phys. Lett., 73A 244 (1979).

15. P.C. Coleman, T.C. Griffith, G.R. Heyland and T.L. Killeen 4th Int. Conf. on Positron Annihilation, Helsingør, Denmark, Abstr, A13 (1976)

16. S.C. Sharma, J.D. McNutt, A. Eftekhari and Y.J.Ataiiyan, Can. J. Phys. 60:610 (1982).

17. J.D. McNutt, S.C. Sharma, M.H. Franklin and M.A. Woodall, Phys. Rev., 20A:357 (1979).

18. O.E. Mogensen, 4th Int. Conf. on Positron Annihilation. Helsingør, Denmark, Abstr. R10 (1976)

19. F.M. Jacobsen, N. Gee and G.R. Freeman, "Positron Annihi-lation", Edited by P.G. Coleman, S.C. Sharma and L.M. Diana, North-Holland Publ. Co., Amsterdam (1982) 92.

20. D.M. Schrader and R.E. Svetic, Can.J. Phys. 60:517 (1982).

21. F.M. Jacobsen, O.E. Mogensen and G. Trumpy, Chem. Phys., 69:71 (1982).

22. O.E. Mogensen and F.M. Jacobsen, Chem. Phys., 73:223 (1982).

23. P. Jansen and O.E. Mogensen, Chem. Phys. 25:75 (1977).

24. B. Levay and O.E. Mogensen, Chem. Phys., 53:131 (1980).

25. G. Wikander and O.E. Mogensen, Chem. Phys., 72:407 (1982).

26. F.M. Jacobsen, O.E. Mogensen and M. Eldrup, Chem. Phys., 50:393 (1980)

27. F.M. Jacobsen, to be published.

28. M. Eldrup, A. Vehanen, P.J. Schultz and K.G. Lynn to be published.

29. T.C. Griffith et al, see these proceedings.

30. J.M. Warman and M.C. Sauer Jr. J. Chem. Phys., 62:1971 (1975).

31. T.S. Stein and W.E. Kauppila, Adv. in At. and Mol. Phys. 18:53 (1982).

32. L. Onsager, Phys. Rev., 54:554 (1938).

33. T.C. Griffith, M. Charlton, G. Clark, G.R. Heyland and G.L. Wright, "Positron Annihilation", Edited by P.G. Coleman, S.C. Sharma, and L.M. Diana, North-Holland Publ. Co., Amsterdam, (1982), 61.

34. T.C. Griffith, private communication.

POSITRON LIFETIME SPECTRA FOR GASES

G.R. Heyland

Department of Physics and Astronomy
University College London
Gower Street, London WC1E 6BT

Observations on positron lifetime spectra for krypton and
xenon and for mixtures of these gases with helium, neon and some
molecular gases are presented, including temperature dependence.
Some of the complex features of the lifetime spectra for CO_2 and
SF_6 are indicated.

1. Introduction

There is no doubt that the major experimental effort directed
to the study of the interaction of slow positrons with matter in
general will be concentrated on the development and application of
intense slow positron beams. These beams should soon find application to the interaction of slow positrons with gases, which is the
special field that concerns us here. However, what might now be
described as the classical method, namely the lifetime technique,
has not yet run out of steam but continues to provide new and
stimulating data.

As the interval between positron conferences has been reduced
to one year, this article inevitably contains material taken from
recent reviews and especially those from the positron group at
University College London. Currently the most important, intriguing
and potentially illuminating interaction must be positronium formation. The latest positronium formation cross sections, Griffith
(1983), for different gases show considerable variety and structure.
In general it is confirmed that the cross sections are large, that
is, comparable with or even larger than the elastic cross sections.
The formation threshold, E_{Ps}, seems to be in the right place and

for most gases the initial rise is very sharp with the cross
section reaching a maximum below the ionisation threshold E_I. A

plateau region extends well beyond E_I and in some cases reveals a

second maximum. One question which concerns us here is – does our
improved knowledge of these cross sections shed any light on, or
suggest modifications to, our interpretation of the lifetime spectra
for gases? At our first meeting, held at York University, Toronto,
Heyland et al (1982) discussed the general nature of positron lifetime spectra for gases. A number of parameters were defined to which
numerical values could be assigned from the data analysis and this
nomenclature will be followed here.

2. Krypton and Xenon and mixtures with He, Ne, H_2, D_2 and N_2

In an attempt to explain the two fast components and the
dearth of ortho-positronium in the spectra of both krypton and
xenon, reported by Heyland et al (1982), further experiments have
been conducted using gas mixtures. The original intention was to
see what effect small quantities of He, Ne and molecular gases would
have on the positronium fraction. Some results for pure krypton
were quoted by Griffith et al (1982), where it was shown that of the
total gas events for krypton at 4.96 amagats, 17% and 16% respectively were due to the two fast components, 56% to the free positron component and 11% to ortho-positronium. The addition of small
quantities of helium had no measurable effect on the number of free
positron events but rapidly increased the number of ortho-positronium events from 11% to 30% at 50% He concentration. The addition
of the helium correspondingly reduced the number of events in the
fast components as well as rapidly increasing both annihilation
rates so that at 15% He concentration neither fast component could
be resolved. These results for Kr – He mixtures also clarify those
previously reported by Griffith and Heyland (1978), Charlton et al
(1979) where values for the positronium formation cross sections
were deduced. The present results confirm that positronium is
formed in pure Kr but that a large fraction of it is rapidly
quenched, presumably from a bound state. The formation of this
bound state is inhibited by small quantities of helium so that the
measured positronium fraction increases. At 50% helium concentration the positronium fraction reaches a plateau at $F = 0.39$ which
is maintained to 98% helium, so that this value of F represents the
true positronium formation fraction for krypton. The contribution
to F from collisions between positrons and helium atoms does not
become significant until the helium concentration exceeds 98% and
this confirms that the mean positronium formation cross section for
helium is only $\sim$ 2% of the total inelastic cross section for krypton
in the energy range 17.8 to 24.6 eV (E_{Ps} to E_I for He). These

effects were less pronounced for Kr – Ne mixtures possibly because

the heavier neon atoms are less effective in moderating the
positronium through the critical energy region for bound state
formation.

Similar results were obtained for xenon mixtures at 297 K.
Figure 1 shows the increase in the ortho-positronium component for
Xe - He mixtures as a function of the helium density for a constant
xenon density of 4.7 amagats. At low helium concentrations the
number of O-Ps events detected increases linearly with the helium
concentration. For helium concentrations greater than 90%, the
positronium fraction attains the value F = 0.51 which is the true
positronium fraction for xenon. The addition of N_2, D_2 and H_2 to
xencn also increased the ortho-positronium component but for much
lower concentrations of the molecular gas, as shown in Figure 2,
with nitrogen the least effective and hydrogen the most effective.
For these gases it is possibly rotational excitation which is
mainly responsible for the effect.

A striking and unexpected effect was the considerable increase
in the equilibrium free annihilation rate at 297 K for the xenon
mixtures which was not observed for the krypton mixtures. On the
assumption that the annihilation rate for the mixture is given by

$$\bar{\lambda}_{mix} = \bar{\lambda}_{Xe} + \bar{\lambda}' \tag{1}$$

where $\bar{\lambda}'$ is the contribution from the added gas calculated from
its partial density ρ' and known $\bar{Z}_{eff} = Z'$ for the pure gas (i.e.
$\bar{\lambda}' = 0.201 \, Z' \, \rho' \, \mu sec^{-1}$), the $\bar{Z}_{eff}$ for xenon rises from its value
of 320 for the pure gas to ~ 420 for the mixtures for very low
concentrations of the added gas and where $\bar{\lambda}'$ is negligible. Data
for He, H_2, D_2 and N_2 are given in Figure 3.

Data for Xe - He mixture containing 8.4% H_2 were also taken as
a function of the xenon partial density; Figure 4. As the density
decreased from 37 amagats, $\bar{Z}_{eff}$ increased steadily from ~ 160
reaching ~ 460 at 6 amagats. At lower densities $\bar{Z}_{eff}$ decreased
approaching the value 320 as for the pure gas. At low densities
it would seem that the increase in $\bar{Z}_{eff}$ for Xe due to the H_2 is
a function of the H_2 partial density rather than the percentage
concentration. Also shown are the positronium fraction F and
$_1Z_{eff}$ parameter for Xe derived from the O-Ps annihilation rate.

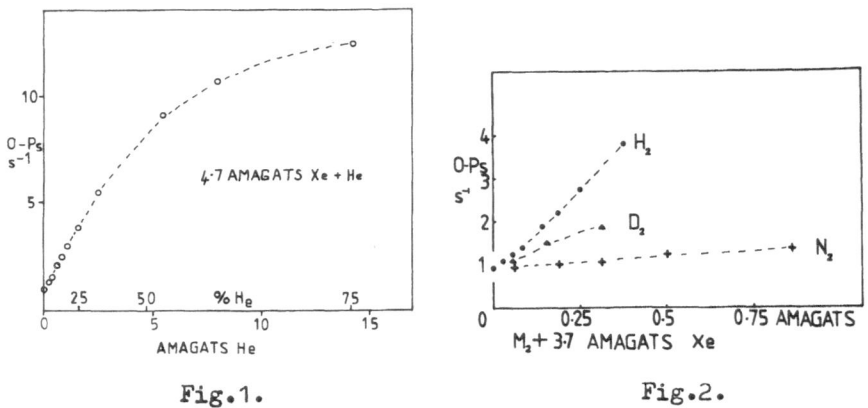

Fig.1.   Fig.2.

Fig.1. O-Ps event rate for Xe-He mixtures.
Fig.2. O-Ps event rate for Xe-molecular gas mixtures.

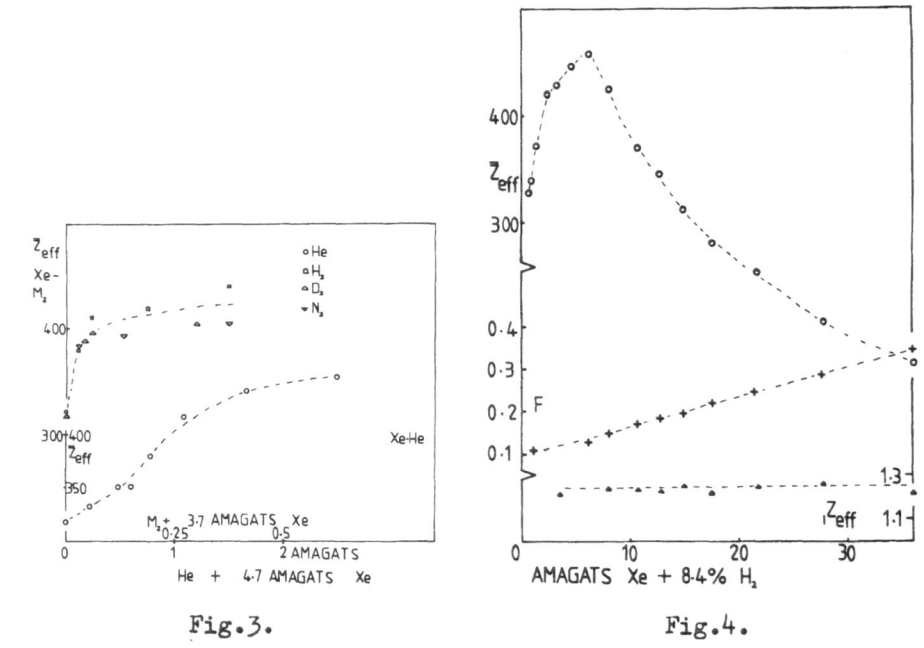

Fig.3.   Fig.4.

Fig.3. $\bar{Z}_{eff}$ for Xe for Xe-He and for Xe-M_2 mixtures.
Fig.4. $\bar{Z}_{eff}$, F and $_1Z_{eff}$ for Xe + 8.4% H_2 mixture as a
        function of Xe density.

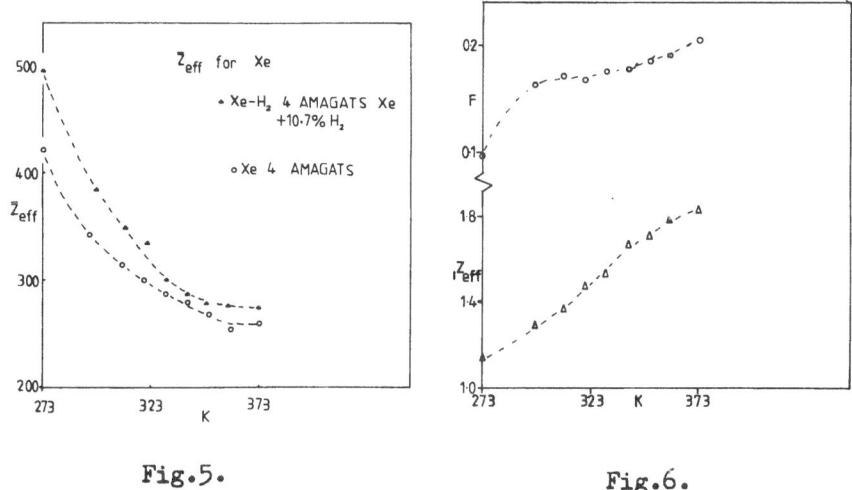

Fig.5.   Fig.6.

Fig.5. Dependence of $\bar{Z}_{eff}$ on temperature for pure Xe and for
      Xe-H$_2$ mixture.
Fig.6. Dependence of F and $_1Z_{eff}$ on temperature for Xe-H$_2$ mixture
      as in Fig.5.

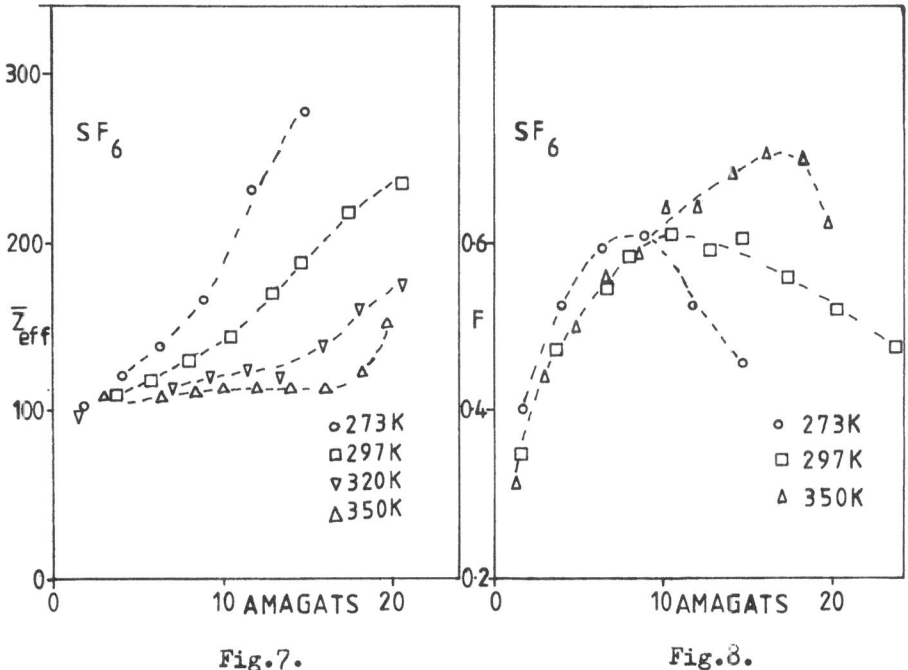

Fig.7.   Fig.8.

Fig.7. Dependence of $\bar{Z}_{eff}$ on density and temperature for SF$_6$
Fig.8. Dependence of F on density and temperature for SF$_6$

3. The Effect of Temperature on Xenon and Xenon-H$_2$ Mixture

The high $\bar{Z}_{eff}$'s for Xe may partly be due to the onset of cluster formation and so be strongly temperature dependent. Lifetime spectra for Xe at 3.7 amagats for the pure gas and for the addition of 10.7% of H$_2$ were taken for the temperature range 273 - 373 K. Figure 5 shows the variation of the equilibrium $\bar{Z}_{eff}$ for Xe as a function of temperature for the pure gas and for the Xe - H$_2$ mixture using equation (1). At 297 K the temperature coefficient of $\bar{Z}_{eff}$ for the pure gas is 2.4 K^{-1} which is several times the value calculated by McEachran et al (1980). The higher values for the mixtures are maintained over the measured temperature range and it is noted that at the higher temperatures the values of $\bar{Z}_{eff}$ for the pure gas approach the calculated values. This behaviour of the $\bar{Z}_{eff}$ parameter for Xenon has yet to receive a satisfactory explanation. For pure gas only about 1% of the free positrons survive to reach the 'equilibrium' region from which the $\bar{Z}_{eff}$ value is determined and it may be that thermal equilibrium is not attained. The addition of a molecular gas will certainly promote more rapid thermalisation and it may also broaden the positron energy distribution during the slowing down process due to rotational excitations.

In Figure 6, values for the positronium fraction F and the ortho-positronium quenching parameter $_1Z_{eff}$ are shown as a function of temperature for the Xe - H$_2$ mixture. The initial considerable decrease in the positronium fraction F on cooling the mixture from 297 K to 273 K is unlikely to be due to contamination, whereas part of the increase in F for the higher temperatures may be due to this cause. The large increase of the $_1Z_{eff}$ parameter for the mixture, which was not observed for the pure gas, was unlikely to be due to contamination. No explanation is offered for these effects.

4. The Molecular Gases CO$_2$ and SF$_6$

The behaviour of the lifetime parameters for the molecular gases is now being studied by several groups. The general complexity of the subject has been indicated in recent reviews. Griffith et al (1982) referred to the close correspondence between $\bar{Z}_{eff}$ and the polarisability. Heyland et al (1982) pointed out that both the $\bar{Z}_{eff}$ parameter and the positronium fraction F were strongly density dependent for many gases. In particular, for CO$_2$ and SF$_6$ at 297 K, the $\bar{Z}_{eff}$'s show an approximately linear increase with

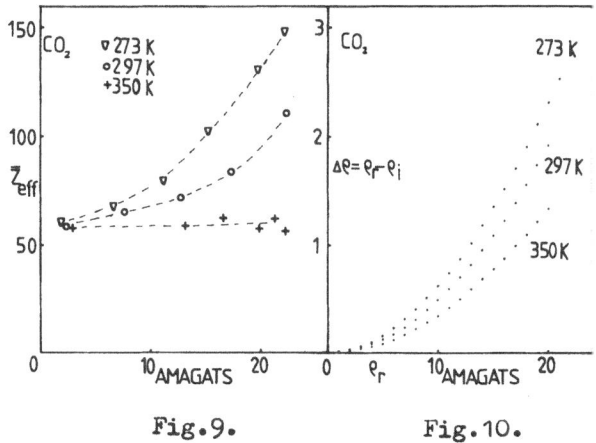

Fig.9.   Fig.10.

Fig.9. The dependence of $\bar{Z}_{eff}$ on density and temperature
       for CO_2.

Fig.10. Density deviations from ideal gas behaviour for CO_2
        for the same temperatures as in Fig.9.

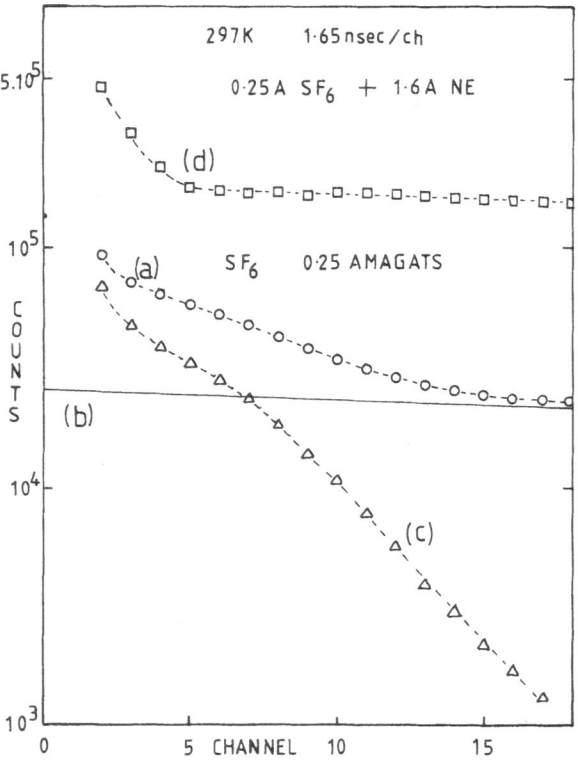

Fig.11. Lifetime data for SF_6 and SF_6 - Ne mixtures.
(a) Restored signal for SF_6 at o.25 amagats. (b) combined
free positron and O-Ps component   (c) Fast component.
(d). Restored signal for 0.25 amagats SF_6 + 1.6 amagats Ne.

density with the values doubling over the range 0 to 20 amagats. The dependence of $\bar{Z}_{eff}$ and F on temperature and density for SF_6 are shown in Figures 7 and 8. The behaviour of $\bar{Z}_{eff}$ for CO_2 is shown in Figure 9. For comparison Figure 10 shows the density deviations from ideal gas law for the same temperature values which are also of similar magnitude for SF_6. For these two gases the steady increase in $\bar{Z}_{eff}$ from very low densities could be explained in terms of an increased $\bar{Z}_{eff}$ for pairs of molecules in close proximity encountered by the positron. These two gases have another feature in common. At very low densities, the spectra for both gases reveal fast components which also appear to have pronounced shoulder regions.

Data for SF_6 at 297 K and 0.25 amagats are shown in Figure 11. Curve (a) is the restored signal in the region of the prompt peak with curve (b) the combined free positron and ortho-positronium components, $(\bar{\lambda}_p \sim \bar{\lambda}_f)$. Curve (c) = (a) - (b) is attributed to a fast component. This fast component with 'shoulder' can be easily observed for densities as low as 0.06 amagats. The addition of 1.6 amagats neon to the 0.25 amagats SF_6 virtually removes this fast component. For the pure gas this fast component disappears as the density is increased above 0.6 amagats, but in the range 2 to 4 amagats two other fast components are observed which appear very similar to those observed in krypton and xenon. Once again, no interpretation of these effects is offered.

In conclusion, the fact that the positronium formation cross sections are still appreciable for positron energies well above the threshold for ionisation has to be taken into account in the interpretation of lifetime spectra for gases. There is little doubt that some energetic positronium will be formed for positron energies extending to $\sim$ 100 eV. However, above the excitation theshold, positronium formation competes with other inelastic processes which become progressively more dominant. Energetic positronium formed with kinetic energy in excess of the binding energy (6.8 eV) is presumed to dissociate in a subsequent collision with the release of a low energy positron. This compound process is equivalent to ionisation and does not invalidate the Ore model. However, it may allow the positronium fraction to increase if stabilisation of energetic positronium can occur with increasing gas density.

Acknowledgements

The work described in this article has been funded by the SERC and has been carried out in collaboration with Dr. M. Charlton, Professor TC Griffith and Dr. GL Wright. The author wishes to

thank the organising committee of the conference on positron scattering in gases held between 19 - 23 July, 1983 at Royal Holloway
College, Egham, Surrey, UK.

References

1:    TC Griffith 'Positronium formation cross sections in various
      gases' This conference.

2:    GR Heyland, M Charlton, TC Griffith and GL Wright 'Positron
      lifetime spectra for gases' Can. J. Phys. $\underline{60}$:471 (1982).

3:    TC Griffith, M. Charlton, G Clark, GR Heyland and GL Wright
      'Positrons in gases - A progress report' Proc. 6th Int. Conf.
      on positron annihilation, Univ. Texas at Arlington (1982) p.61.

4:    TC Griffith and GR Heyland 'Experimental aspects of the study
      of the interaction of low-energy positrons with gases' Phys.
      Rep. $\underline{39C}$ (1978) p.258.

5:    M Charlton, TC Griffith, GR Heyland and KS Lines "Cross sections
      for positronium formation in gases' J. Phys. B. $\underline{12}$:L 633(1979).

6:    RP McEachran, AG Stauffer and LEM Campbell 'Positron scattering
      from krypton and xenon' J. Phys. B. $\underline{13}$:1281 (1980).

THE CALCULATION OF POSITRON LIFETIMES

D. M. Schrader

Chemistry Department
Marquette University
Milwaukee, WI 53233 USA

INTRODUCTION

A recent admirable review by Drachman,[1] given on the occasion of the first conference in this series, contains a helpful section on lifetime calculations. Drachman reviews two interesting techniques for improving the accuracy of calculated lifetimes which are based not upon calculating better wave functions but rather upon operator transformations. In the present review we will concentrate on the former approach: the improvement of wave function accuracy.

The calculation of accurate wave functions for many-electron systems now constitutes an advanced technology, and one which is still growing rapidly. Although the application of this technology to mixed electron-positron systems is straightforward, only a little work has been carried out with it.[2] Most authors use the conventional Hartree-Fock method in spite of its well-documented deficiencies for lifetime calculations.[3] In the next few pages, we will briefly review lifetime calculations in relation to several widely used methods for wave function calculation, with emphasis on the method of correlated configurations which is, in our view, most promising for providing accurate lifetimes in the near future.

THE HARTREE-FOCK METHOD

The Hartree-Fock wave function for a one-positron, closed shell n-electron system has the form

$$\Psi=\phi_p(r_p)\alpha(p)A[\phi_1(r_1)\alpha(1)\phi_1(r_2)\beta(2)\cdots\phi_{\frac{n}{2}}(r_n)\beta(n)]. \tag{1}$$

The spin function chosen for the positron is arbitrary, and A antisymmetrizes and normalizes. The Hartree-Fock equations are, in atomic units.

$$
\begin{aligned}
&[-\tfrac{1}{2}\nabla^2-\tfrac{z}{r} + \sum_j(2J_j-K_j)-J_p-\varepsilon_i]\phi_1=0,\\
&[\ \tfrac{1}{2}\nabla^2+\tfrac{z}{r} - \sum_j(2J_j) \qquad -\varepsilon_p]\phi_p=0.
\end{aligned}
\tag{2}
$$

The symbols are familiar and have obvious meanings. This system of equations was first written down by Chang Lee a quarter century ago.[4] The annihilation rate for the wave function (1) is given by

$$\lambda\alpha\left\langle\Psi|\sum_{j=1}^{n}\delta^3(r_{jp})|\Psi\right\rangle. \tag{3}$$

The integral can be written

$$2\sum_{j=1}^{n/2}\int|\phi_p(r)|^2|\phi_j(r)|^2dr. \tag{4}$$

We emphasize that (3) is not true in general. One sometimes sees comments in the literature to the effect that the annihilation rate is simply the expectation value of a sum of delta functions, and that the correct expression is (4), where ϕ_p and ϕ_j are termed the "wave functions" of the positron and the electron, resp. This inaccurate language should be avoided. A whole system has a wave function, not the individual particles which constitute it. The ϕ's above are properly termed "orbitals". The deceptively simple expression (3) is correct only for special forms of the wave function Ψ, such as (1). A more general expression for λ is given below.

When the Hartree-Fock method is used to calculate annihilation rates, the results are too small by up to an order of magnitude.[5] The annihilation rate is a much more exacting property to calculate than the momentum distribution. This has been identified with the role of electron-positron correlation.[6] Hartree-Fock calculations[5] of the angular correlation of the annihilation photons for the positronium halides are in perfect agreement with laboratory measurements.[7]

The Hartree-Fock method is occasionally used for scattering calculations. The distinctions between bound and scattering calculations in eq(2) lie in the sign of the value of ε_p and in the normalization condition for ϕ_p. The Hartree-Fock method does not account for target polarization, which is a long-range correlation effect. Consequently, unsatisfactory phase shifts result unless the polarization potential is overshadowed by a stronger term, as is the case if the target possesses a permanent dipole moment or a charge. In the absence of such terms, the polarizabilities/phase shifts and the annihilation rates are seen to be long-range and

short-range manifestations of the same effect: positron-electron
correlation, and qualitatively incorrect results are to be expected
from the Hartree-Fock method.

The simple forms (3) and (4) are preserved with a more rigorous
method, which might be termed the "Correlated Hartree-Fock" method.[8]
In this method, the functional appearance of the wave function is
similar to that given in eq(1), except that the electronic orbitals
depend upon the positron position as well as on the positions of
one electron: $\phi_j \equiv \phi_j(r_e, r_p)$. Then the argument list of ϕ_j in (4)
is: r,r. This method permits the introduction into the wave
function of electron-positron correlation, but the equations which
the electronic orbitals ϕ_j must satisfy are much more difficult to
solve than eqs(2). For this reason, the correlated Hartree-Fock
method has not been widely used.

The polarized orbital method[9] considerably predates the correlated Hartree-Fock method, but it can be understood as a series of
approximations on the latter.[10] It has been widely used for positron-
atom and -molecule scattering, most notably by the York group. When
applied to noble gas targets, it gives a very respectable account
of both polarizabilities[11] and annihilation rates.[12] A necessary
step in comparing calculated annihilation rates with observations
is averaging over the velocity of the positron.[10] This is so because
at room temperature, the velocity dependent annihilation rates of
most systems which have been considered so far vary markedly over
the width of the velocity distribution function.

THE CONFIGURATION INTERACTION METHOD

Calculation on positronium hydride, PsH, indicate that the
conventional configuration interaction method and its scattering
counterpart, the close coupling method, probably are not good choices
for positron-atom calculations. Ludwig and Parr[13] calculated a nine-
term CI wave function for PsH with a basis set featuring functions
with nonintegral principal quantum numbers. They included four
configurations of the type $(ss' + s's)s''$ (the positronic function
is the last one listed), three of the type $(sl + ls)l'$ with $l=p,d$,
and f, and two more with correlated electrons: p^2s and d^2s. Their
energy, -0.7502 au, is only 95.2% of the accurate value.[14] Wave
functions of comparable flexibility give 99.8% of the energy for
helium.[15] The contrast of the expectation values of $\delta^3(r_{12})$ and
$\delta^3(r_{1p})$ is even more stark--a one-term ss' wave function for He gives
a value of $\delta^3(r_{12})$ 55% too high,[16] but Ludwig and Parr's nine-term
CI function is less accurate, giving a value of $\delta^3(r_{1p})$ 61% too
low.[14] The data in Table I, taken from the above references, summarizes our observations (the parenthetical numbers are the percentages
of the calculated values of the "true" values). Although the wave
functions and systems are not exactly comparable, it is seen that

Table I. Comparison of purely electronic and mixed systems

CI wave function	He E	$\langle\delta^3(r_{12})\rangle$	H- E	$\langle\delta^3(r_{12})\rangle$	PsH E	$\langle\delta^3(r_{1p})\rangle$
ss'	-2.8757 (99.0)	0.1648 (155)	-0.5133 (97.3)	0.00538 (196)	-0.6452 (81.9)	
$4ss'$	-2.8789 (99.1)	0.1767 (166)			-0.6833 (86.8)	
$ss' + 3l^2$	-2.8974 (99.8)				-0.7502 (95.2)	0.00812 (39)

PsH, the archetypical positronic atom, presents conventional CI with a most uncooperative system, both from the standpoint of its energy and, equally important for positronic atoms, $\langle\delta^3(r_{1p})\rangle$, which is proportional to the annihilation rate. In contrast, Neamtan's[17] very simple correlated PsH wave function,

$$e^{-\alpha r_1 - \beta r_2 - \gamma r_{1p} - \delta r_{2p}} + (1\leftrightarrow 2),$$

gives 96.2% of the energy and 96% of $\langle\delta^3(r_{1p})\rangle$. The examples cited above are from rather old literature; CI calculations with thousands of terms are being carried out today. CI calculations will ultimately converge, of course. The point of Table I is not that the CI method is inaccurate, but that its convergence rate is poor for positronic systems.

THE METHOD OF CORRELATED CONFIGURATIONS

As formulated by Sims and Hagstrom,[18] the method of correlated configurations consists, in essence, of variationally combining configurations, each of which contains a correlating factor for a selected pair of particles. In the rest of the contribution, we limit our discussion to the example system e^+Li for the sake of concreteness.

In the correlated configuration interaction method, the wave function is expressed as a sum of configurations,

$$\Psi(X_1,X_2,X_3,X_p) = \sum_K C_K \Phi_K(X_1,X_2,X_3,X_p). \tag{5}$$

e^+Li has four particles, with space and spin coordinates X_1, X_2, etc. Particles 1, 2, and 3 are electrons and p denotes a positron. Each configuration Φ_K is a projected product of a spin factor (discussed below),

$$\chi_1 = \tfrac{1}{2}(\alpha_1\beta_2 - \beta_1\alpha_2)(\alpha_3\beta_p - \beta_3\alpha_p), \tag{6}$$

and a spatial factor

$$F_K(r_1,r_2,r_3,r_p) = r_{ij}^{\nu_K} e^{-a_K r_{ij}} \prod_{s=1,\frac{1}{2},3,p} \phi_{\kappa s}(r_s). \tag{7}$$

The projection operator (not shown) is a product of an orbital angular momentum projection operator and an electron antisymmetrizer. The one-particle function ϕ is a Slater-type orbital.

The exponential factor $e^{-a_K r_{ij}}$ in eq(7) is not a part of the original formulation.[18] The new integrals which arise on account of this factor may be evaluated by an application of the generalized Laplace expansion, and present no essential difficulties.[19] We believe that this factor is necessary for those positronic atoms for which the lowest threshold includes positronium, as is the case for e^+Li. Since we believe the ground state of e^+Li has zero orbital angular momentum, the dominant configuration will contain the spatial factor

$$F_1 = (e^{-0.5r_{3p}})(1s_{2.7}(r_1)1s_{2.7}(r_2))(n_3s_{\zeta_3}(r_3)n_p s_{\zeta_p}(r_p)). \tag{8}$$

The first parenthetical factor on the right side represents virtual positronium; the second, the core; and the third, the valence electron and the positron. We can simulate positronium at any desired distance by approximately choosing values of the unspecified parameters in the third parenthetical factor above. Other significant configurations are those which correlate the core electrons.

Clary[20] has studied e^+Li with this method, except he always took a_K in eq(7) to be zero; i.e., no exponential correlating factor was used -- only linear and quadratic factors of an interparticle coordinate were used. Clary's calculation was very ambitious: 84 configurations. The correlated pairs include all the possibilities: valence electron-positron, core-core electron, core-valence electron, and core electron-positron.

With such an extensive wave function, convergence to the correct energy should be expected to within a small fraction of an electron volt. However, the final energy given, which seems to be converged to within ~0.003 eV, is 0.558 eV above the Ps + Li^+ threshold. A more modest wave function representing Ps and Li^+ at a large separation will give the threshold correctly to within ~0.002 eV: one need only combine Hylleraas'[21] eight-term correlated wave function for Li^+ with the known Ps wave function, and let the distance between them become large. Clary's suggests that he is converging on a resonance, but his wave function and its convergence properties exhibit none of the characteristics one associates with such states.[22] Sims and Hagstrom's famous calculation on Be is comparable:[18] the same method was used, and the number of configurations was similar. Their

calculated energy for Be was only 0.008 eV above the experimental
value corrected for relativistic effects. Similarly, a calculation
on Li$^-$ gives an energy 0.011 eV too high.[23] The apparent discrepancy
in Clary's calculation is not yet understood.

The Spin Problem

The ground state of e$^+$Li certainly has $S_e=\frac{1}{2}$ and (as already
noted) zero orbital angular momentum. It follows that the states
^{1,2}S and ^{3,2}S (using the notation ^{a,b}L, where a=2S + 1, and b=2S$_e$ + 1,
S and S_e being the total electron-plus-positron spin and S_e the
electron spin[6]) are candidates for the ground state of the system.
These states will turn out to be very close in energy: the corresponding states of positronium are separated by only 8.4 X 10^{-4} eV
due to continuum coupling and the spin-spin interaction, the singlet
state being lower. For e$^+$Li, spin polarization of the core is another
source of splitting, and this will probably turn out to be a larger
effect than continuum coupling and the spin-spin interaction. We
expect, however, that the complete fine structure interval for the
e$^+$Li ground state will nevertheless be a very small fraction of an
electron volt.

The spin function χ_1, eq(6), has S=0 and $S_e=\frac{1}{2}$, and so qualifies
as a possible spin factor of the wave function of the ^{1,2}S state.
Another such function, linearly independent of χ_1, is

$$\chi_2=\tfrac{1}{2}(\alpha_1\beta_p-\beta_1\alpha_p)(\alpha_2\beta_3-\beta_2\alpha_3). \tag{9}$$

This function can be included in the trial wave function implicitly
by using χ_1 in conjunction with the spatial factor $P_{13}F_K$.[18]

χ_1 and χ_2 span the singlet spin space for this system. The
triplet spin space is spanned by three spin functions, two with
$S_e=\frac{1}{2}$ and one with $S_e=\frac{3}{2}$. The latter will give rise to the terms
^{3,4}L which will be many electron volts above the ground state. We
need consider only the two others, the dominant of which is, for
$S_Z=1$,

$$\chi_3=\tfrac{1}{\sqrt{2}}(\alpha_1\beta_2-\beta_1\alpha_2)\alpha_3\alpha_p, \tag{10}$$

and another, which we can represent as

$$\chi_4=\tfrac{1}{\sqrt{2}}(\alpha_1\beta_3-\beta_1\alpha_3)\alpha_2\alpha_p. \tag{11}$$

But $\chi_4=P_{23}\chi_3$ so, again, we need use only χ_3 in our calculations,
and we can include the effects of χ_4 in our trial wave function by
using appropriate spatial factors.

The Calculation of Annihilation Parameters

A complete quantum field theoretic of this problem does not appear to be in the literature.[4],[24] It is possible, however, to deduce effective operators which appear to be generalizations of eq(3) for wave functions of arbitrary form. As before, we write Ψ as the pre-annihilation wave function of one positron and n electrons, but now the functional form of Ψ is unspecified. For calculating the 2γ annihilation rate, we want to project out of Ψ the component which corresponds to a relative spin singlet for all possible positron-electron pairs. We also need to project out a resultant linear momentum component for the annihilating particles, and a wave function of the $n-1$ spectator electrons. Finally, we must have coalescence of the annihilating pair, so we can write

$$\gamma_j(k) = \left\langle \Psi \,\middle|\, \sum_{\mu=1}^{n} \delta^3(r_{\mu p}) \sigma_{00}(\mu, p) e^{ik \cdot R_{\mu p}} \Phi_j(\mu) \right\rangle. \tag{12}$$

This expression is a generalization of one due to Neamtan, Darewych, and Oczkowski.[17] The sum is over electrons and the integration is over the space and spin coordinates of the $n + 1$ particles in the parent state wave function Ψ. The symbol $\Phi_j(\mu)$ stands for the wave function of the j-th state of the $(n-1)$ electron daughter, with the j-th electron absent. $R_{\mu p}$ is $\frac{1}{2}(r_\mu + r_p)$, the position of the center of mass of the annihilating particles, $\sigma_{00}(\mu, p)$ is the singlet spin factor for the annihilating particles, and k is the resultant momentum of the two photons. The probability for 2γ annihilation with the photons having resultant momentum k is

$$N_{2\gamma}(k) \alpha \sum_j |\gamma_j(k)|^2. \tag{13}$$

The sum is over daughter states. Using closure, the right side can be written as an expectation integral involving only the parent wave function:[25]

$$N_{2\gamma}(k) \alpha \left\langle \Psi \,\middle|\, \sum_{\mu=1}^{n} \Omega_\mu \,\middle|\, \Psi \right\rangle. \tag{14}$$

Ω_μ is a projection operator involving only the annihilating particles:

$$\Omega_\mu = \left| \delta^3(r_{\mu p}) \sigma_{00}(\mu, p) e^{ik \cdot R_{\mu p}} \right\rangle \left\langle e^{ik \cdot R_{\mu p}} \sigma_{00}(\mu, p) \delta^3(r_{\mu p}) \right|. \tag{15}$$

To use these expressions on a system without a nucleus such as Ps or the positronium negative ion ($e^-_2 e^+$, or Ps$^-$), one must confine the system in a big volume V which cancels out in the subsequent integrations. To get the correct 2γ annihilation rate

$$\frac{r_0^2 c}{2a_0^3} \approx 8.0 \times 10^9 \text{sec}^{-1},$$

the missing proportionality constant in (13) and (14) must be

$$\frac{r_0^2 c}{2\pi^2 a_0^3}$$

where r_0 is the classical radius of the electron.

Returning to our illustrative example system, e^+Li, we denote $\tilde{F}_K$ as the angular momentum-projected spatial factor F_K in eq(7). If we first integrate over the spins of the annihilation particles, then over r_p and k, we arrive at integrals containing the delta functions $\delta^3(r_\mu - r_{\mu'})$, $\mu = 1, 2,$ and 3. Integrations over $r_{\mu'}$ and the remaining spin variables yield for the $^{1,2}S$ state

$$
\begin{aligned}
^1\lambda_{2\gamma} = \frac{4\pi r_0^2 c}{a_0^3} \Sigma_K \Sigma_{K'} C_K^{(1)*} C_{K'}^{(1)} \{ & \langle \tilde{F}_K(1233) | \tilde{F}_{K'}(1233) + \tilde{F}_{K'}(2133) \\
& -\tfrac{1}{2}(\tilde{F}_{K'}(3123) + \tilde{F}_{K'}(2313) + \tilde{F}_{K'}(3213) + \tilde{F}_{K'}(1323)) \rangle \\
& -\tfrac{1}{2}\langle \tilde{F}_K(1232) | \tilde{F}_{K'}(3122) + \tilde{F}_{K'}(1322) \qquad (16) \\
& -\tfrac{1}{2}(\tilde{F}_{K'}(1232) + \tilde{F}_{K'}(2312) + \tilde{F}_{K'}(3212) + \tilde{F}_{K'}(2132)) \rangle \\
& -\tfrac{1}{2}\langle \tilde{F}_K(1231) | \tilde{F}_{K'}(2311) + \tilde{F}_{K'}(3211) \\
& -\tfrac{1}{2}(\tilde{F}_{K'}(1231) + \tilde{F}_{K'}(3121) + \tilde{F}_{K'}(2131) + \tilde{F}_{K'}(1321)) \rangle \},
\end{aligned}
$$

where now the integrations are over the spatial coordinates of the three electrons only. Similarly, the 2γ annihilation rate for the $^{3,2}S$ state is the much smaller quantity

$$
\begin{aligned}
^3\lambda_{2\gamma} = \frac{\pi r_0^2 c}{a_0^3} \Sigma_K \Sigma_{K'} C_K^{(3)*} C_{K'}^{(3)} \{ & \langle \tilde{F}_K(1231) | \tilde{F}_{K'}(1231) - \tilde{F}_{K'}(3121) \\
& + \tilde{F}_{K'}(2131) - \tilde{F}_{K'}(1321) \rangle + \langle \tilde{F}_K(1232) | \tilde{F}_{K'}(1232) \qquad (17) \\
& - \tilde{F}_{K'}(2312) - \tilde{F}_{K'}(3212) + \tilde{F}_{K'}(2132) \rangle \}.
\end{aligned}
$$

Clearly these results are not gotten by simply calculating the expectation value of a delta function over the parent wave function, as is sometimes assumed. The parenthetical superscripts on the C_K's in eqs(16) and (17) remind us that the secular problem will not yield quite the same linear coefficients for the singlet and triplet states of our system. The spin dependence of the two outer particles will have the effect of slightly uncoupling the two core electrons through exchange in the case of the triplet but not the singlet. This differential core polarization will be a very small fraction of the total energy, but it might be the dominant contribution to the singlet-triplet splitting. The lifetime of "ortho" e^+Li in a magnetic field is a sensitive function of this splitting, so it is essential to calculate it accurately.

SECOND ORDER CALCULATION OF POSITRON LIFETIMES

We use the Ritz variational principal to calculate an approximate wave function Ψ:

$$\Psi \text{ from } \min\langle\Psi|H|\Psi\rangle = E_{min}, \langle\Psi|\Psi\rangle = 1. \tag{18}$$

This Ψ may be of high quality but it is not exact except for positronium, and its error $\varepsilon\Psi$ is unknown. Fortunately, the corresponding error in E_{min} is second order in ε. Unfortunately, when the annihilation rate is calculated with this Ψ,

$$\lambda = \langle\Psi|\Omega|\Psi\rangle, \Omega = \Sigma_\mu\Omega_\mu, \tag{19}$$

first order errors result. In this section we will briefly review some very old mathematical physics which shows how to reduce the error in λ to second order.[26]

The operator Ω does not perturb the system, but we can write a formal first-order equation nonetheless:

$$(H^{(0)} - E^{(0)})\Psi^{(1)} + (\Omega - \bar{\Omega})\Psi = 0$$

$$\text{Re}(\langle\Psi^{(1)}|\Psi^{(0)}\rangle) = 0 \tag{20}$$

where the bar above means averaging over Ψ, and where

$$(H^{(0)} - E^{(0)})\Psi = 0 \tag{21}$$

The solution $\Psi^{(1)}$ has no particular physical significance. If we express it in terms of another unknown,

$$\Psi^{(1)} = F\Psi, \tag{22}$$

then eq(19) may be written

$$([T,F] + \Omega - \bar{\Omega})\Psi = 0 \tag{23}$$

Where T is the kinetic energy part of H. Thus we have eliminated $H^{(0)}$ and $E^{(0)}$, which are not ordinarily extracted in the Ritz procedure anyway, and we have also removed the orthogonality requirement shown as the second of eqs(20). The annihilation rate is now given up to order ε^2 as

$$\lambda = \langle\Psi|\Omega|\Psi\rangle + \langle\Psi|\{F, H - E_{min}\}|\Psi\rangle, \tag{24}$$

where curly brackets denote the anticommutator. The direct solution of eq(22) is not convenient, but a variational determination of F is achieved by minimizing the functional

$$\left\langle \Psi \middle| [F,[H,F]]+2F(\Omega-\bar{\Omega}) \middle| \Psi \right\rangle. \tag{25}$$

Only the exact solution of eq(22) will guarantee second order errors in λ, eq(24), but an accurate variational determination of F will enable one to approach that ideal as closely as one desires.

The near degeneracy of many positronic systems such as e^+Li is not a problem for us because these states have different quantum numbers which are preserved by our "perturbation" Ω. The substitution (22) fixes the nodes of $\Psi^{(1)}$ to coincide with those of Ψ, but this is probably not significant.

REFERENCES

1. R. J. Drachman, Can. J. Phys. 60, 494 (1982).

2. J. R. Winick and W. P. Reinhardt, Phys. Rev. A 18, 910, 925 (1978); J. W. Humberston, Can. J. Phys. 60, 591 (1982); G. D. Doolen, J. Nuttall, and C. J. Wherry, Phys. Rev. Lett. 40, 313 (1978); L. T. Choo, M. C. Crocker, and J. Nuttall, J. Phys. B 11, 1313 (1978).

3. D. M. Schrader in "Positronium and Muonium Chemsitry", H. J. Ache, ed., (Advances in Chemistry Series 175, Am. Chem. Soc., 1979), pp. 203ff.

4. Chang Lee, Zh. Eksp. Teor. Fiz. 33, 365 (1957).

5. A. Farazdel and P. E. Cade, J. Chem. Phys. 66, 2612 (1977).

6. C. F. Lebeda and D. M. Schrader, Phys. Rev. 178, 24 (1969).

7. O. E. Mogensen, Chem. Phys. 37, 139 (1979); P. Pfluger, K.-P. Ackermann, R. Lapka, E. Schupfer, R. Jeker, H.-J. Guntherodt, E. Cartier, and F. Heinrich, Synthetic Metals 2, 285 (1980).

8. D. M. Schrader, Phys. Rev. A 1, 1070 (1970).

9. A. Temkin, Phys. Rev. 107, 1004 (1957).

10. D. M. Schrader and R. E. Svetic, Can. J. Phys. 60, 517 (1982).

11. R. P. McEachran, A. G. Ryman, and A. D. Stauffer, J. Phys. B 10, L681 (1977); R. P. McEachran, A. D. Stauffer, and S. Greita, J. Phys. B 12, 3119 (1979).

12. R. P. McEachran, D. L. Morgan, A. G. Ryman, and A. D. Stauffer,
     J. Phys. B 10, 663 (1977); R. P. McEachran, A. G. Ryman, and A.
     D. Stauffer, J. Phys. B 11, 551 (1978); R. P. McEachran, A. G.
     Ryman, and A. D. Stauffer, J. Phys. B 12, 1031 (1979); R. P.
     McEachran, A. D. Stauffer, and L. E. M. Campbell, J. Phys. B 13,
     1281 (1980).

13. O. G. Ludwig and R. G. Parr, Theoret. chim. Acta 2, 440 (1966).

14. Y. K. Ho, Phys. Rev. A 17, 1675 (1978).

15. G. R. Taylor and R. G. Parr, Proc. Natl. Acad. Sci. 38, 154
     (1952).

16. J. N. Silverman, O. Platas, and F. A. Matsen, J. Chem. Phys. 32,
     1402 (1960); D. P. Chong and D. M. Schrader, Mol. Phys. 16, 137
     (1969); C. L. Pekeris, Phys. Rev. 115, 1216 (1959).

17. S. M. Neamtan, G. Darewych, and G. Oczkowski, Phys. Rev. 126, 193
     (1962).

18. J. S. Sims and S. A. Hagstrom, Phys. Rev. A 4, 908 (1971); J. S.
     Sims and S. A. Hagstrom, J. Chem. Phys. 55, 4699 (1971).

19. S. Hagstrom, private communication (1982).

20. D. C. Clary, J. Phys. B 9, 3115 (1976).

21. E. A. Hylleraas, Z. Phys. 65, 209 (1930).

22. A. Temkin, ed. "Autoionization" (Mono Book Corp., Baltimore,
     (1966).

23. J. S. Sims, S. A. Hagstrom, D. Munch, and C. F. Bunge, Phys. Rev.
     A 13, 560 (1976).

24. R. A. Ferrell, Rev. Mod. Phys. 28, 308 (1956); O. E. Mogensen,
     "Fourth International Conference on Positron Annihilation,
     Helsingør, Denmark, 23-26 August 1976," G3.

25. D. M. Schrader, "Positron Annihilation," P. G. Coleman, S. C.
     Sharma, and L. M. Diana, eds. (North-Holland, 1982), p. 71.

26. A. Dalgarno and J. T. Lewis, Proc. Roy. Soc. (London) A233, 70
     (1955); C. Schwartz, Ann. Phys. (N. Y.) 6, 170 (1959); G. G.
     Hall, Adv. Quantum Chem. 1, 241 (1964); J. O. Hirschfelder, W.
     B. Brown, and S. T. Epstein, Adv. Quantum Chem. 1, 255 (1964).

TECHNIQUES FOR STUDYING SYSTEMS CONTAINING MANY POSITRONS

A.P. Mills, Jr.

Bell Laboratories
Murray Hill
New Jersey 07974

I INTRODUCTION

We are at the threshold of being able to study systems containing finite amounts of antimatter: the electron-positron plasma, positronium molecules and droplets, and surfaces having comparable electron and positron densities. The necessary ingredients for such studies are well known.[1] One must first obtain 10^{-9} sec bursts containing $\sim 10^7$ slow positrons each either from a pulsed electron accelerator [2-4] or from a strong reactor-produced ^{64}Cu radioactive source [5] combined with time bunching stages. [6] The positron bursts must then be brought to focus on a few hundred angstrom diameter spot on a target surface by means of repeated stages of acceleration, focusing and moderation (brightness enhancement [7]). While no one has made such a positron source, progress is being reported on all aspects of the problem.

The purpose of this workshop and this paper in particular is to stimulate further thought on the subject of intense positron beams. We have been encouraged to think especially about the future development of these beams. The next section gives a brief look at some new areas of study that might become available with advanced beam technology and a discussion of terminology: intensity, flux, brightness, phase space density, fluidity, Debye screening, and space charge limitation. Section III is a brief description of our best radioactive source and moderator. Section IV gives some details about an existing time bunching apparatus and points out a way to bunch bunches. The last section is about obtaining one positron at a time.

II Intensity and Brightness

Beams of monoenergetic low energy positrons have been available for more than a decade. Many interesting experiments have been performed that have increased our understanding of positronium atomic physics, positron-molecule scattering, and positron surface interactions. The understanding gained in the latter studies has enabled us to improve our techniques for making slow positrons. There are now beams in several laboratories that may be termed "intense" because they have passed the threshold at which one should be shielded from the annihilation radiation. These beams with intensities I up to $I = 5 \times 10^6$ slow e^+ sec^{-1} are the equivalent of a 270μ Ci source of 511 keV γ-rays. In combination with various timing, polarizing, focusing, deflecting, detecting and transporting techniques, these beams have given adequate counting rates in many different investigations. In the near future even more intensity will be available with the perfection of the electron accelerator and nuclear reactor sources. A beam of $I = 10^9$ e^+/sec could be called "very intense" because at any instant there would be typically more than one positron in the target. In this sense the LINAC beams and bunched beams are very intense pulsed sources, since they yield bursts having instantaneous intensities $I > 10 e^+$/nsec.

Besides its intensity I, a beam if characterized by its flux or intensity per unit area,

$$\Phi = I/(\pi r^2), \tag{1}$$

where r is the beam radius. A beam may also be characterized by its brightness per unit energy, [8]

$$R = \Phi/(E \sin^2 \theta), \tag{2}$$

where E is the kinetic energy of the beam particles and θ is their angular divergence. One of the most useful parameters describing a swarm of N particles is its density in phase space

$$\rho_\phi = N/\Omega, \tag{3}$$

where Ω is the phase space volume occupied by the swarm. For a beam of particles with energy uniformly distributed between E and $E+\Delta E$, with propagation directions uniformly distributed up to a maximum angle θ with respect to the beam axis and with beam radius r, we have approximately

$$\rho_\phi = R/(2m\Delta E). \tag{4}$$

Liouville's theorem states that ρ_ϕ is constant if the beam is only acted upon by conservative forces. A corollary is that the brightness per unit energy is also conserved if ΔE is a constant. If a beam is focused using a short focal length lens, the maximum obtainable flux satisfies the inequality

$$\Phi_{max} < ER. \tag{5}$$

For a system of unpolarized Fermions, [9] $\rho_\phi \leqslant 2/h^3$. A particle

beam with ρ_ϕ close to this limit would be called nearly degenerate.
A beam with $\Delta E = 1eV$ would thus have a brightness per unit energy
no greater than

$$R_{max} < 4m\Delta E/h^3 = 3.21 \times 10^{28} \text{ sec}^{-1} \text{ cm}^{-2} \text{eV}^{-1}. \quad (6)$$

A 10 keV beam with $\Delta E = 1eV$ therefore has a limiting flux of

$$\Phi_{max} < 4mE\Delta E/h^3 = 3.21 \times 10^{32} \text{ sec}^{-1} \text{ cm}^{-2}. \quad (7)$$

This means that one could in principle have a 1 mA, 10 keV beam
focused to a 1Å spot.

We should now consider thresholds and limits associated with
the internal potential energy of the particle swarm. Suppose a
beam of energy E and a radius a_o is converging with half angle
θ_o towards a focal point. The beam will stop converging at a
diameter a such that the potential energy per unit length

$$U = \sigma^2 \ln(\ell/a) \quad (8)$$

equals the initial transverse kinetic energy per unit length

$$T = (\sigma E/q) \sin^2\theta. \quad (9)$$

Here σ is the charge per unit length $\sigma = qI\sqrt{m/2E}$, q is the particle
charge, m is the particle mass, I is the beam intensity and ℓ is
roughly the length of the beam, $\ell \gg a$. Conservation of energy
implies

$$\sigma\ln(\ell/a) + (E/q)\sin^2\theta = \text{const.} \quad (10)$$

We thus have [10]

$$a_{min} = a_o \exp\left\{-E\sin^2\theta/\sigma q\right\}. \quad (11)$$

Space charge effects will only show up when focusing a beam having
$\sin^2\theta < \sigma q/E$. For $I = 10^7$ e$^+$/nsec and $E = 2.5$ keV this means θ must
be greater than about .01 radian. Thus, space charge should not
prevent us from focusing even a 1mA beam to a very small diameter.

A fluid threshold [11-14] having to do with the random motion
or temperature T of a beam occurs when the inter-particle potential
energy $\frac{q^2}{r}$ becomes larger than kT, $\beta^2 \equiv e^2/(kTr) > 1$. Here r is the
interparticle spacing. A beam will exhibit plasma effects when the
Debye screening length $\lambda_D = r/\beta$ is less than the dimensions of the
beam.

Finally we must consider flux, intensity, and brightness

thresholds having to do with the special properties of positrons aimed at a solid target surface. The first threshold will be at a brightness and intensity suitable for performing low energy positron diffraction or LEPD experiments [15] in a reasonable time with a resolution comparable to LEED. Taking $E = 50$ eV, $I = 10^6 sec^{-1}$, $2r = 1$ mm and $\theta = 0.01$ radian we have $R = 2 \times 10^{10}$ sec^{-1} cm^{-2} eV^{-1}. Since LEPD intensities are about $10^{-3}-10^{-4}$, such a beam would yield 1% precision at 100 different energies in about an hour. Our present beams have more than the required intensity but a brightness per unit energy of only about 10^7 sec^{-1} cm^{-2} eV^{-1}. A single stage of focusing and remoderation (brightness enhancement [7]) would suffice to make such a LEPD beam. [16] The same beam accelerated to several keV would be suitable for imaging defects on a 1 μm scale.

It is energetically possible to form positronium molecules when two surface positrons interact with each other. By a suitable choice of surfaces, the positronium desorption energy can be obtained from the Ps$_2$ binding energy, $\Delta E = 0.20$ eV. [17] Using standard thermodynamic arguments [18] it is easy to show that the Ps$_2$ desorption rate is

$$z_{Ps_2} = (\hbar n n_+/m_+^2)(1-r_{Ps_2})e^{(\Delta E-2E_a)/kT}, \qquad (12)$$

where m_+ is the surface positron effective mass, m is the free electron or positron mass, n_+ is the surface density of positrons, r_{Ps_2} is the thermally averaged Ps$_2$ reflection coefficient and E_a is the activation energy for thermal desorption of surface positrons to form positronium. Equation 12 was derived using the relations $2\mu_{Ps} = \mu_{Ps_2}$ and $\mu_{Ps} = \mu_{e^+} + \mu_{e^-}$ for the chemical potentials of the constituents: a 2D gas of e$^+$ and metallic e$^-$ in 3D in equilibrium with a gaseous mixture of Ps and Ps$_2$. The Ps$_2$ is taken to have spin zero. The remaining steps in the derivation are identical to Eqs. 8-11 of the paper by Chu et al. [18] The ratio of the number of positrons desorbed from the surface as molecules to the number desorbed as atoms [18] is then

$$z_{Ps_2}/z_{Ps} = (n_+/n_0)(1-r_{Ps_2})/(1-r_{Ps})e^{(\Delta E-E_a)/kT} \qquad (13)$$

where the critical surface density n_0 is given by

$$n_0 = 2m_+kT/\pi\hbar^2. \qquad (14)$$

At room temperature (T = 300 K) and with $m_+ = m$, we have $n_0 = 2.5 \times 10^{13}$ cm^{-2}. It is possible to adjust E_a by coating certain surfaces with a partial monolayer of an adsorbate, for instance [19] O$_2$ on Al or Cs on Ni. If we set $E_a = 0.1$eV, then assuming $r_{Ps_2} = r_{Ps}$, a positron surface density $n_+ = 5 \times 10^9$ cm^{-2} gives us a 1% Ps$_2$ formation probability relative to thermal Ps. To see Ps$_2$ molecules we will need a positron flux $\phi = 5 \times 10^{18}$ cm^{-2} sec^{-1} and

a brightness per unit energy $R = \phi/E = 5 \times 10^{14}$ $sec^{-1}cm^{-2}$ eV^{-1}.
About three stages of brightness enhancement would be required.

Other interesting phenomena will occur when the distance
between surface positrons is comparable to a room temperature
positron deBroglie wave length (10 Å) and a positronium Bohr radius
(1Å). Given a sufficient number of positrons at one time ($\sim 10^6$)
we might observe Bose condensation effects and the positron-electron
fluid respectively.[20] One and two more stages of brightness
enhancement and large e^+ pulses are going to be required to get to
this stage. The positron annihilation γ-ray laser [21,22] will be
a possibility if we can compress the interparticle spacing by another
factor of 100 to less than a Compton wavelength. Entirely new
physics may appear if the spacing can be made comparable to the
classical electron radius as suggested by Winterberg.[23] These
last two regimes require energies much exceeding that which we
associate with slow positron beams. Nevertheless, the new techniques for obtaining many positrons with high density and narrowly
defined energies may serve as a starting point for bigger accelerators.

Figure 1 summarizes the above. In general it seems that interesting new physics will appear each time the brightness and instantaneous intensity of our beams both increase by several orders of
magnitude. We have only begun our journey in this intensity vs.
brightness space but the project seems feasible and the motivation
is not lacking.

III Single Crystal Tungsten Moderator

The presently most efficient and brightest slow positron sources
make use of clean flat single crystal metallic moderators.[24] High
energy positrons from β-decay or pair production stop in the moderator
and a small fraction of them diffuse to the surface. If the metal
has a negative affinity for positrons they can be ejected from the
surface. The majority of the positrons are emitted normal to the
surface [25] with an essentially thermal energy spread about $-\phi_+$,
the negative of the positron workfunction.

Single crystal tungsten has been found to be an excellent
positron moderator. [26-28] It can be cleaned by roasting in O_2
to remove carbon followed by gentle heating in vacuum to vaporize
the tungsten oxide. A satisfactory moderator surface also can be
obtained by heating to white heat in vacuum. Figure 2 shows a
mounting arrangement that allows the moderator to be heated by
electron bombardment without melting its holder. The moderator was
heated with 170 W of electron current several times for 20 sec.
each. After several cycles, the pressure in the vacuum vessel
(1×10^{-10} torr) rose to no more than 2×10^{-9} torr during the heating
Using an emissivity of 0.33, the maximum temperature of the crystal
is calculated to be 2900 K. Figure 3 shows how the count rate
changes as the moderator was moved toward the source in the back

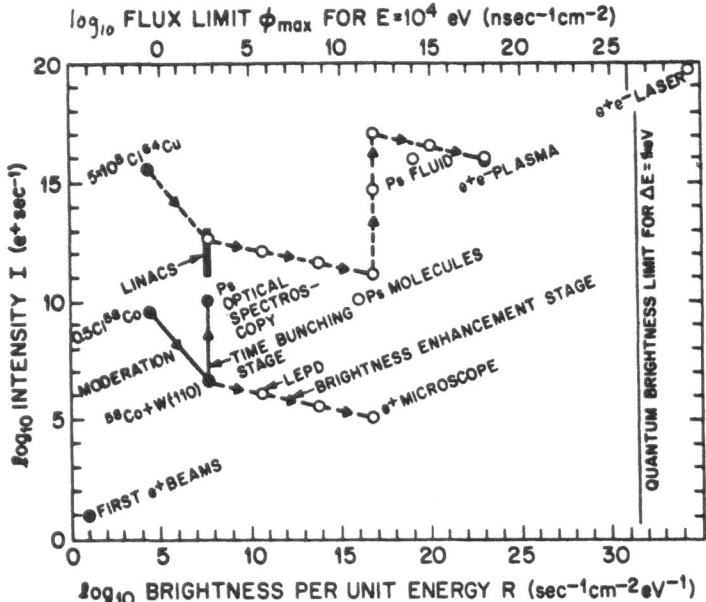

Fig.1   Our present location in e$^+$ intensity vs. brightness space,
        some possible goals for future work and some possible paths
        for getting there through stages of bunching (vertical
        dashed lines) and brightness enhancement (dashed line sloping
        down to the right).

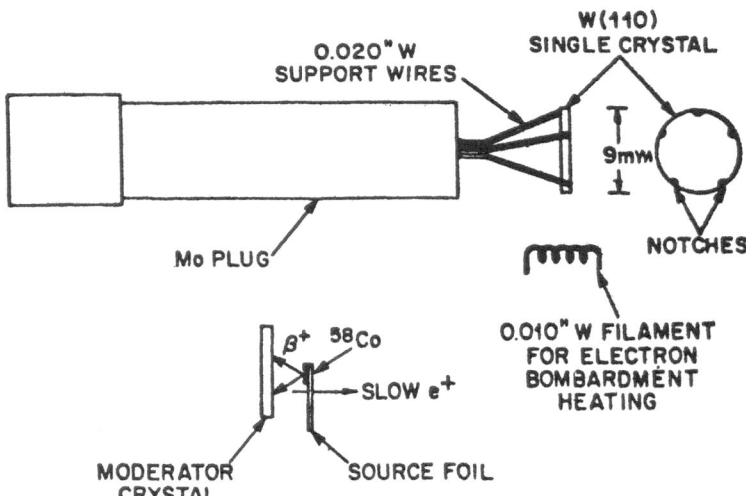

Fig. 2   Mounting arrangements for a tungsten single crystal positron
         moderator.

scattering geometry [29,30] shown in Figure 2. The source is
310 mCi ^{58}Co plated by New England Nuclear Corp. on a copper plated
electro-polished tungsten ribbon 0.005 inches thick by .060 inches
wide. This source preparation [24] helps reduce β^+ absorption by
keeping the ^{58}Co deposit near the surface of the source foil. The
source spot is about 3 mm long. The positron intensity is calculated from the background subtracted count rate of a 3x3 inch NaI(Tl)
detector behind a 2 inch thick 8x8" Pb brick with a 1.5 inch hole
in the center. The face of the counter is 59 cm from a grounded
dirty Si target in the vacuum system. The γ-rays pass through a
glass window about 3/16inch thick. The moderator is biased +7 V
relative to ground. Using 92% as the detector efficiency for
511 keV γ-rays, the fast positron to slow positron conversion
efficiency ε is computed to be $\varepsilon = 0.17\%$ including a 10% absorption
in the glass window. Experience has shown that the count rate
increases by about 10% when the positrons are driven into the target
at keV energies. We infer that $\varepsilon = (0.19 \pm 0.03)\%$. This result
is in agreement with that of the Brookhaven group. [31]

Figure 4 shows retarding field energy spectra obtained with the
same moderator when attempts were being made to get an improved
brightness and yield by coating [28] the W with Cu. The composite
moderator does work slightly better than plain Cu, but the clean W
moderator is best of all and very easy to prepare. When it is biased
to -1.5 V, the positron energy distribution is as narrow as that
from the Cu single crystal moderator [24] ($\sim$ 1/4 eV) and the yield
is better. The 5 eV spread of longitudinal energies with the moderator at +7 V is the image of the 2.5 eV W work function after the
positrons have entered the 150 Gauss transport magnetic field. The
field at the source is about 75 Gauss.

IV Time Bunching Accelerators

Given the constraints of Liouville's theorem, it is possible
to increase the instantaneous intensity of a slow positron beam by
trading energy resolution for spatial density. [6] Figure 5 shows
the overall view of a time bunching accelerator for a magnetically
transported slow positron beam. The positrons are first trapped
in a magnetic bottle by increasing their transverse energy spread.
The positrons pass through an rf cavity shown in Figure 6 tuned to
the cyclotron resonance frequency, about 408 MHz. The Q of the
cavity is $\sim$ 50 and it is typically driven with $\sim$ 50 mW of rf power.
The capacitor plates are 1 inch apart and 2.5 inches wide. The
parallel inductance is provided by the vertical plates and the
6 inch diameter vacuum can. The input impedance is 50 Ω and no
impedance matching element (double stub tuner) is required. It is
important that no insulator be used between the plates because they
can easily charge up and deflect the beam.

The positrons are dumped out of the magnetic bottle by an
accelerator section shown in Figure 5. The accelerator is constructed of reentrant rings (shown in detail in Figure 7) so that

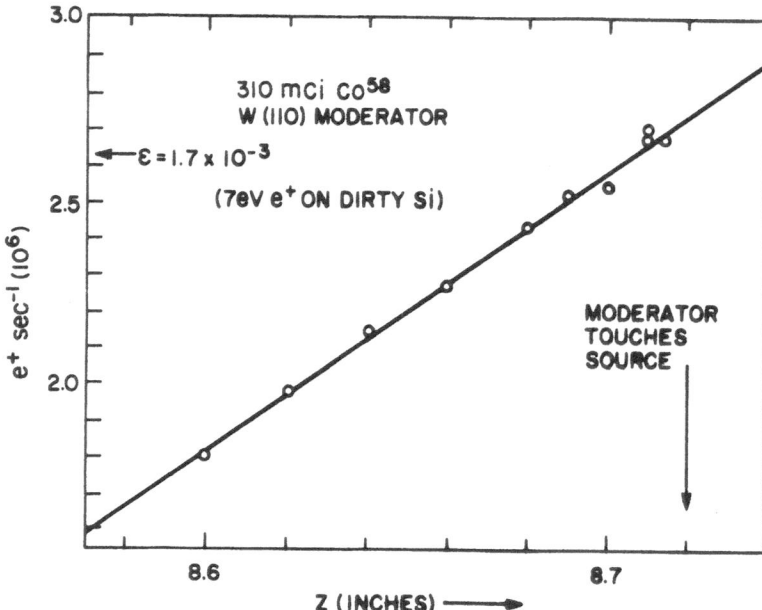

Fig. 3 Slow e$^+$ count rate vs. moderator position.

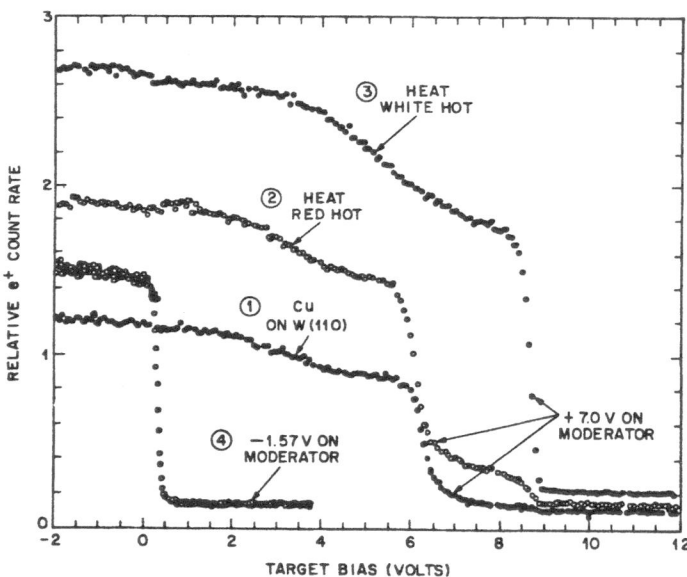

Fig. 4 Slow e$^+$ energy spectra from 1) W(110) heated to white heat
and then coated with Cu; 2) the same heated red hot to anneal
the Cu; 3) the same heated white hot to remove the Cu; and 4)
the same reverse biased to give a very mono-energetic beam.

the supporting Al_2O_3 rods are not exposed to the positron beam.

Between the rings are resistors the values of which are chosen to make a quadratic potential well when a high voltage pulse is applied to one end of the accelerator. The positrons initially have low velocities (< 1eV) in the accelerator and thus all arrive at the minimum of the quadratic potential after one quarter of a simple harmonic oscillator period. The target is placed at this minimum.

The accelerator is divided into two sections, an active part 83 cm long and a passive part 66 cm long, in order to keep the positron pulse from having a wide time spread. Ring S_{53} of Figure 7 is biased slightly positive (about 6 volts) and thus closes one end of the magnetic bottle. The passive part of the accelerator is biased negatively at the Au grid shown in Figure 7. The quadratic potential is established by resistors between accelerator rings $S_{54} - S_{67}$ whose values are R, 3R, 5R, ... 23R with R = 30 K Ω. The resistors are ultra high vacuum compatible carbon film resistors with Ni leads.

The active part of the accelerator is designed to continue the quadratic potential and is actually a distributed 50 Ω attenuator. The series resistors between the accelerator rings are parallel combinations of 1/8 watt carbon resistors. The tinned copper leads are spot welded to the stainless rings. Every fifth or tenth ring is connected to ground through a parallel combination of similar resistors. Figure 8 shows the resistor values used. The series resistances are evenly distributed between the accelerator rings. The accelerator matches 50 Ω very well, as demonstrated by Figure 9 which shows the reflections one sees from the high voltage pulse input when a fast rise time pulse from a sampling scope is applied.

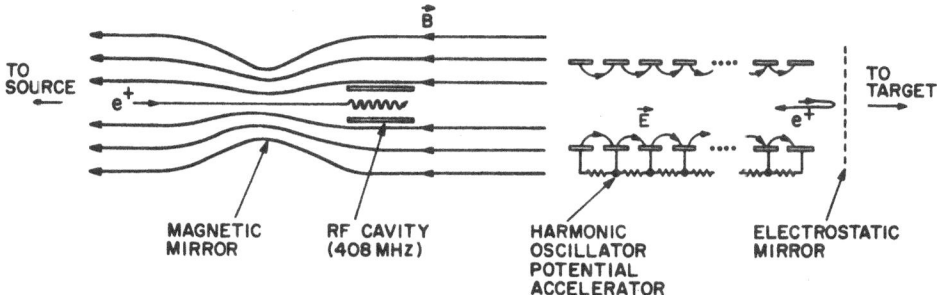

Fig. 5   Time bunching accelerator showing the magnetic bottle and
         quadratic potential accelerator.

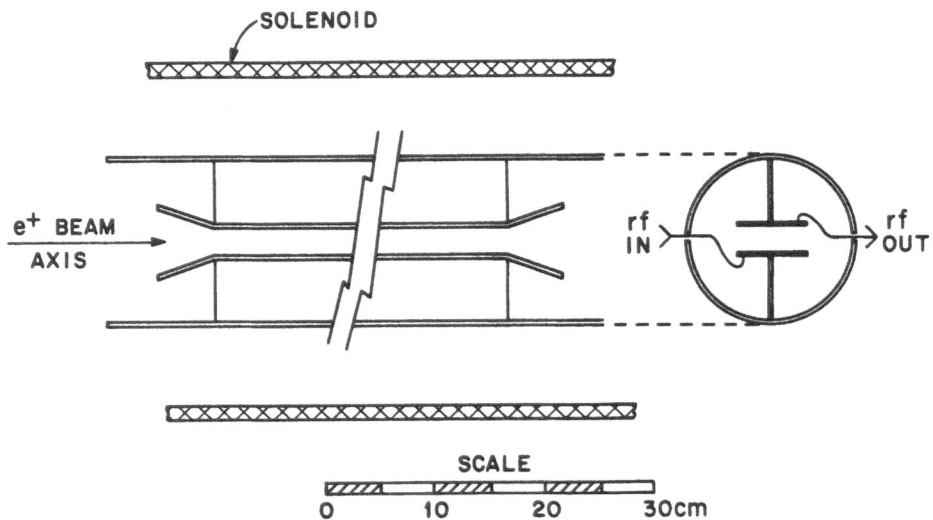

Fig. 6 RF cavity for exciting cyclotron oscillations to trap positrons in the magnetic bottle.

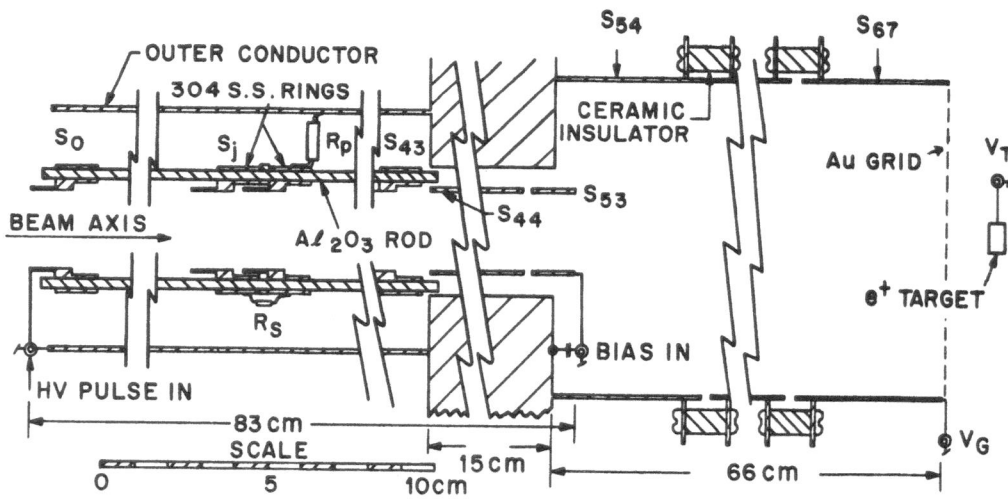

Fig. 7 Quadratic potential accelerator details.

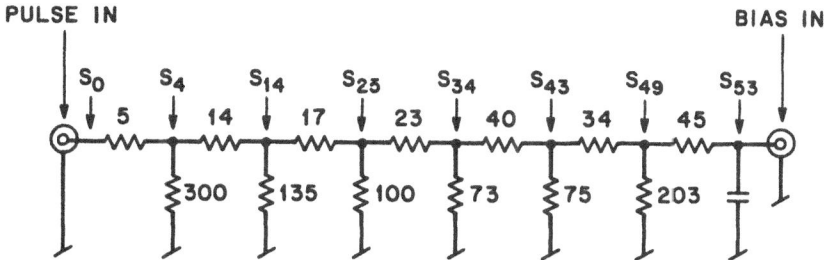

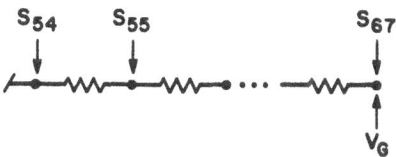

Fig. 8 Resistance values for the quadratic potential accelerator.

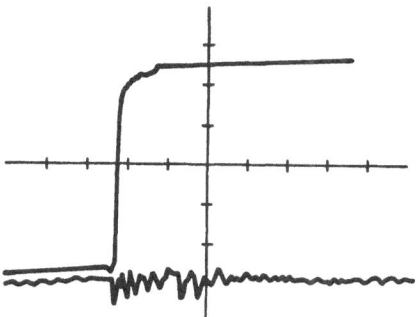

Fig. 9 Reflections (lower trace) observed when a fast pulse
        (upper trace) is applied to the accelerator input. The time
        scale (horizontal) is 2nsec/division.

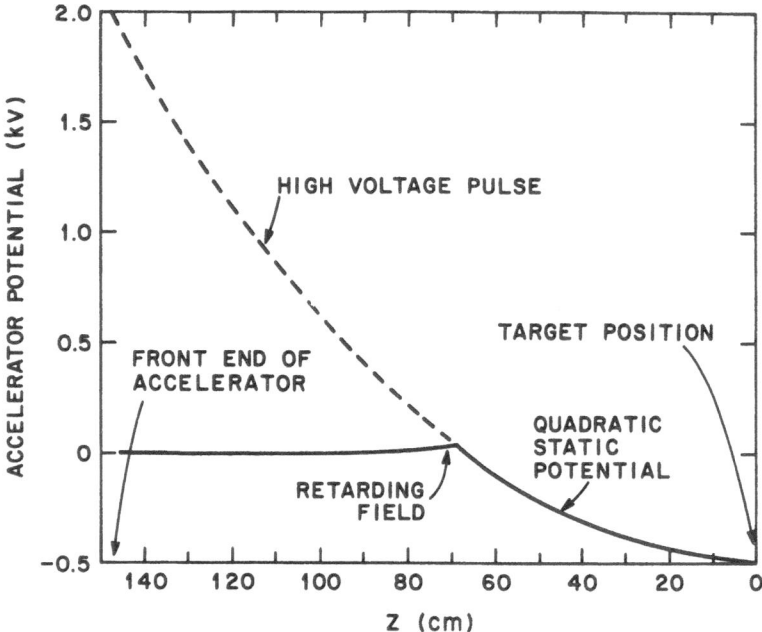

Fig. 10 Potentials on the accelerator. Solid line: accumulation
        stage; dashed line: acceleration stage.

     Figure 10 shows the potentials on the accelerator in the
accumulation stage (solid line) and in the bunching stage (dashed
line). The high voltage pulse is obtained from an HY2 thyratron
operated with a floating cathode and triggered by an avalanche
transistor as shown in Figure 11. Filament power is supplied through
a high frequency transformer. The output pulse delay time is 50
nsec and its rise time 7 nsec 10%-90%. Using the flux of slow positrons from a 300 mCi ^{58}Co source and W(110) moderator we obtain
8 nsec FWHM pulses of 100 ± 20 positrons each at a 1 kHz repetition
rate limited by heating of the resistors in vacuum. When the rf
trap is turned off the bunches become 32 times less intense.

     Figure 12 shows a time spectrum obtained with this accelerator.
The Cu target was biased + 100 V relative to the Au grid in Figure
7. Some of the positrons are reemitted from the Cu ($\sim$ 30%), pass
through the grid, go up the harmonic oscillator potential, and back
to the grid. Evidently a transient associated with the accelerator
pulse makes the Cu target potential slightly more negative at the
moment of emission so that the returning positrons are repelled
back through the Au grid. From the slope of the curve we deduce

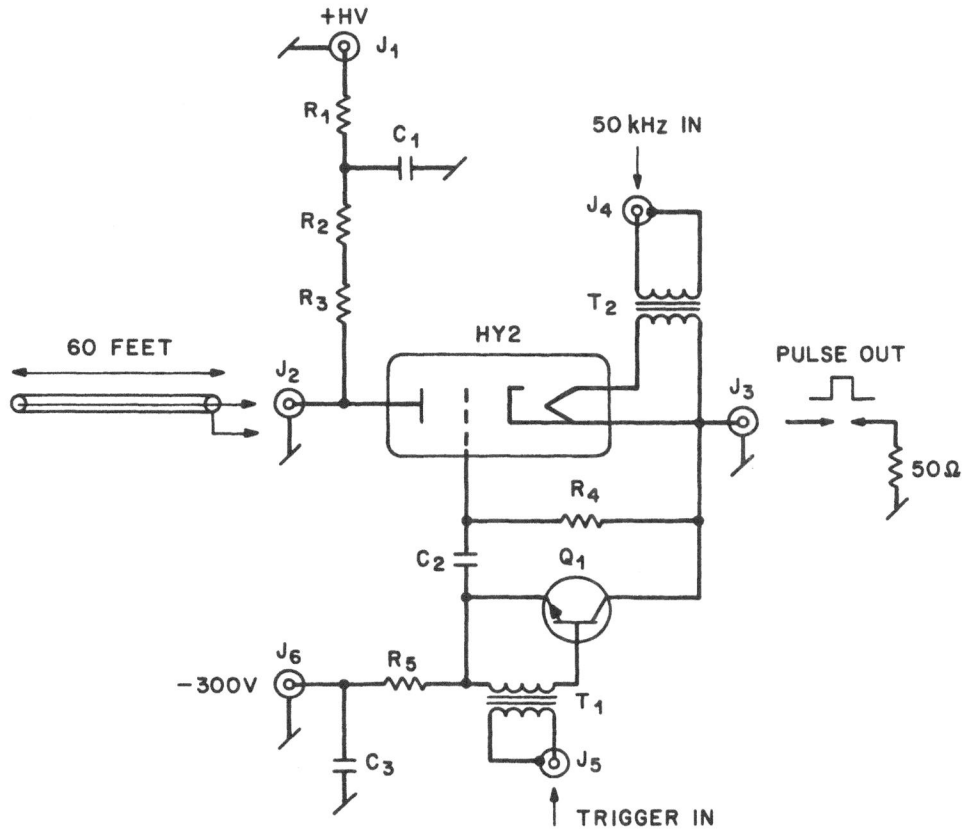

Fig. 11 Circuit diagram for the Thyratron high voltage pulser that drives the accelerator. Q_1 = 2N5271; R_1 = 50kΩ; R_3 = 1.5kΩ; R_4 = 100kΩ. C_1 = 5nF; C_2 = 8.25pF; C_3 = 10nF; T_1 = 5 turns: 10 turns on 3/8" 4C4 ferrite toroid; T_2 = 32 turns no. 22: 9 turns no. 18 on 1/2" 3D3 ferrite toroid. There is a layer of mylar tape insulation between the two windings on the transformers. The 50 kHz rf power supplied to J_4 is adjusted to give 6.3 V across the thyratron filament after it has warmed up.

that about 8% of the positrons annihilate on each pass through the nominally 95% transmitting grid. Since we are able to observe about 40 bounces the positrons must be very mono-energetic or the well quite harmonic.

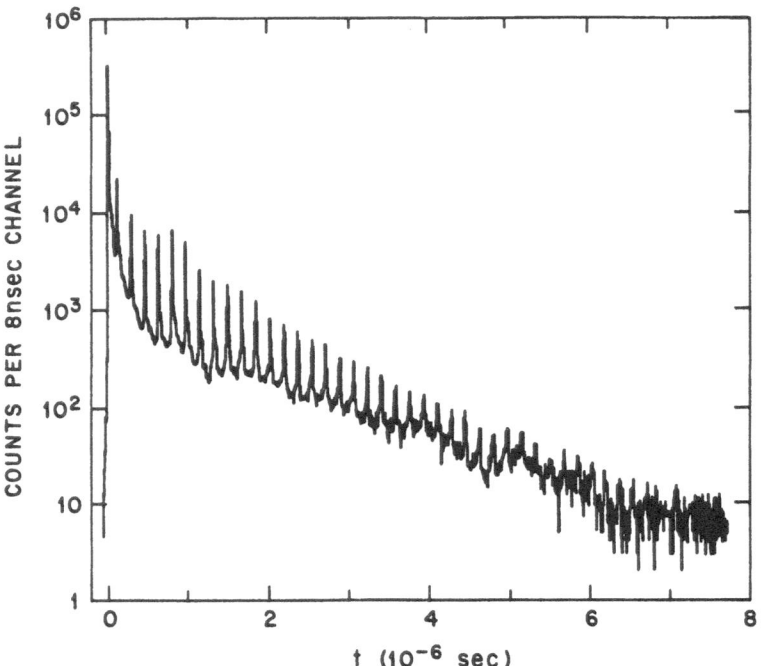

Fig. 12   Annihilation time spectrum obtained with the magnetic bottle
and quadratic potential time bunching accelerator.

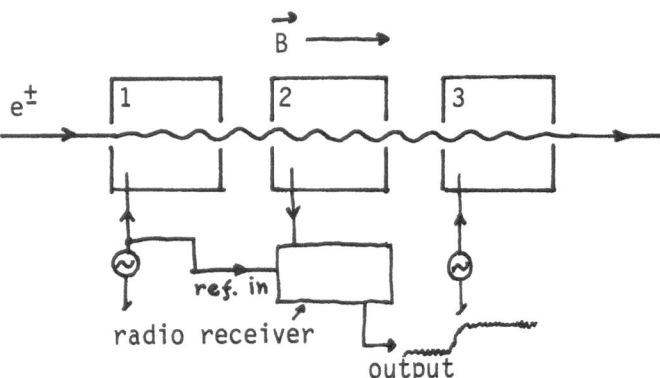

Fig. 13   Non-destructive cyclotron motion detector for slow electrons
and positrons.

Figure 12 suggests that we should be able to collect many positron bunches to make a super bunch. Each bunch would be re-moderated on a W or Cu single crystal moderator surface and stored in a zero potential region of length ℓ in front of the moderator. We must of course cut a hole in the grid so the positrons will not annihilate. Every time a positron bunch arrives the moderator potential is made to dip a few volts below ground so the reemitted positrons will get trapped in the zero potential region. Only a finite storage efficiency is possible because some of the previously stored positrons will be kicked out of the trap if they are close to the target when its potential dips. If the positrons have a velocity v (6×10^7 cm sec^{-1} for 1 eV e$^+$) they travel a distance vΔt during the pulse of time Δt. For Δt = 10nsec, vΔt = 6 mm. Since vΔt/ℓ of the positrons are eliminated during each pulse, one can store a maximum of ℓ/vΔt pulses in this way. Setting ℓ = 60 cm we have ℓ/vΔt = 100. When the trap is filled we could transfer the positrons to a second quadratic potential accelerator. We could then expect to obtain 10 super bunches per second, each with more than 10 positrons. It is obvious that pulsed electron accelerators [2-4] provide superior sources of bunched positrons. Nevertheless, when the experiments contemplated in Section II are attempted we may need multiple bunching of even these sources.

V. Single Positron Sources

In beautiful experiments at the University of Washington, Dehmelt, Schwinberg and Van Dyck have captured single electrons and positrons in a Penning trap and cooled them to liquid He temperatures. [32] They have measured precisely the properties of single particles such as their masses and magnetic moments. In a parallel development Malmberg and colleagues at La Jolla have made single component (non neutral) electron plasmas, [33] essentially a many particle version of the Dehmelt et al. Penning trap. Their goal is to observe plasma effects and eventually to observe the liquid and Wigner crystal states at low temperature.[34,35] The Stanford group working with Fairbank has an interest in dropping single electrons and positrons to determine their gravitational interaction. They have suggested [36] a method for cooling positrons to very low temperatures.

As an extension of these ideas, we can think of shrinking Malmberg's plasma to a line with our goal being to obtain a positron beam in which the particles are never closer in time than a certain amount. We could then do experiments free of the effects of pile up. For a random source of intensity I sec^{-1}, the piled up fraction of the count rate is p = 2I Δt ϵ, where Δt is the pile-up resolution time or the coincidence resolving time and ϵ is the counter efficiency including solid angle. If p_o is the maximum acceptable pile up fraction we must have I < p_o/2$\epsilon\Delta$t. If the counter efficiency is

high, $\varepsilon > p_0/2$, the counting rate in an experiment can be improved
by a factor of $2\varepsilon/p_0$ and pile up totally eliminated by using a
uniform source of intensity $I = 1/\Delta t$. Experiments in which such a
source would be useful include lifetime measurements in which p_0
is small and ε large, and searches for rare events. [37]

We can think of making a steady beam as follows. Suppose we
have a 30 K positron or electron beam confined to a spatial region
of radius a = 10 μm by a 100 kG magnetic field B. The beam travels
through tiny high Q TE mode microwave cavities tuned to the cyclotron resonance frequency ω_c = eB/mc and having cold resistive damping elements at a temperature T. Then according to O'Neill [35]
the particles lose energy at a maximum rate $\Gamma = (\pi e^2/mNV)^{\frac{1}{2}}$ where N
is the number of particles in the cavity of volume V. This expres-
sion holds if the collision rate is less than the damping rate
ω_c/Q of the cavity mode. At 100 kG, ω_c = 2 × 10^{12} sec^{-1}. Using
cavities 1mm long and 1mm in diameter and assuming n = 100cm^{-1} we
would have a damping rate of Γ = 3 × 10^5 sec^{-1}. The longitudinal
motion of the particles is transferred to transverse motion at a
similar rate by Rutherford scattering. The whole beam moving with
its original mean velocity v = 3 × 10^6 cm sec^{-1} would thus internally thermalize [38] to a temperature T in a length of a few times
v/Γ = 10cm. If kT is less than e^2/a = 1 K the particles will
separate longitudinally and can be extracted from the last of the
damping cavities one at a time at a rate nv = 3 × 10^8 sec^{-1}. Such
a beam of electrons or positrons would have very good energy resolution useful for studying vibrational energy losses, etc.

Simply observing such an "anti-bunched" beam would be of interest. If the little microwave cavities also have a TM mode tuned
to the electron spin resonance frequency, the beam will become
highly polarized as the spins relax. The relaxation rate is roughly
10^4 sec^{-1} if Q = 10^3. It is interesting to wonder whether the coupling to the cavities will allow the electron string to freeze or
possibly exhibit pairing.

Another approach to one-at-a-time slow electrons and positrons
is the non-destructive detection of low energy particles. [32]
Lynn [39] has suggested that one could use a super-conducting pick
up coil and squid detector to observe single slow positrons. He
estimates that a 10 μsec averaging time constant would be necessary
to eliminate spurious noise events. A second method shown in Fig.
13 would use three microwave cavities: one to excite the cyclotron
motion of the particle to an energy E_e, a second to sense the cyclotron motion and a third oscillating out of phase with the first one
to take away energy E_e from the cyclotron motion and leave the particle nearly in its ground state again. The signal-to-noise ratio
for a TE$_{111}$ detector cavity with noise temperature T, quality factor
Q, length d and radius R is

$$S/N = 7.0(E_e/kT)(2\pi Q)^2(r_o/d)(1 + \pi^2 R^2/d^2 j_{1,1}^2), \quad (15)$$

where $r_0 = e^2/m$ is the classical electron radius and $j_{1,1} = 1.841 \ldots$ is the first zero of $J_1'(x)$. For $Q = 50,000$, $\omega/2\pi = 1$GHz, $d = 2R$ and $T = 300$K we have a signal-to-noise ratio of 10 with $E_e = 50$eV. This amount of energy yields the same S/N independent of ω because $Q \cong \sqrt{\sigma/\omega}$, where σ is the cavity wall conductivity.

Obviously one needs much smaller cyclotron energies E_e for detection of a particle at cryogenic temperatures. It is interesting to think of the possibility of implementing the Fairbank falling positron experiment using this type of non-destructive detector.

REFERENCES

[1] A.P.Mills, Jr., Science 218,335 (1982)
[2] D.G.Costello, D.E.Groce, D.F.Herring and J.W.McGowan, Phys. Rev. B5 1433 (1972)
[3] R.Howell, R.A. Alvarez and M. Stanek, Appl.Phys.Lett. 40, 751 (1982)
[4] M.Begemann, G.Graff, H.Herminghaus, H.Kalinowsky and R.Ley, Nucl,Instr. and Meth. 201,287 (1982). See also contribution by Graff et al to this workshop.
[5] Work in progress at Brookhaven National Laboratory.
[6] A.P.Mills, Jr., Appl.Phys. 22,273 (1980)
[7] A.P.Mills, Jr., Appl.Phys. 23,189 (1980)
[8] K.F.Canter and A.P.Mills, Jr., Can.J.Phys. 60,551 (1982)
[9] F.W.Sears, Introduction to Thermodynamics; 2nd edition, (Addison-Wesley, Reading, MA. 1953).
[10] J.R.Pierce [Theory and Design of Electron Beams (D.Van Nostrand, New York, 1954) p.147] gives an expression for which $\sin\theta$ is replaced by $\tan\theta$ and E is the longitudinal beam energy.
[11] E.Wigner, Trans. Faraday Soc., 34,678 (1978).
[12] J.H.Malmberg and T.M.O'Neil, Phys.Rev.Lett. 39,1333 (1977).
[13] P.M.Platzman and P.A.Wolff, Waves and Interactions in Solid State Plasmas, (Academic Press, NY, 1973).
[14] L.Spitzer, Physics of Fully Ionized Gases (Wiley, New York, 1962),
[15] I.J.Rosenberg, A.H.Weiss and K.F.Canter, Phys.Rev.Lett. 44, 1139 (1980).
[16] See contribution by K.F.Canter to this volume.
[17] W.F.Brinkman, T.M.Rice and B.Bell, Phys.Rev. B8, 1570 (1973)
[18] S.Chu, A.P.Mills, Jr., and C.A.Murray, Phys.Rev. B23, 2060 (1981); F.Reif, Fundamentals of Statistical and Thermal Physics (McGraw Hill, New York, 1965) p. 324.
[19] K.G.Lynn and D.Gidley, private communication.
[20] The author wishes to thank P.M.Platzman and W.F.Brinkman for discussions.
[21] C.M.Varma, Nature 267,686 (1977); M.Bertolotti and C.Sibilia, Appl.Phys. 19,127 (1979).

[22] R.Ramaty, J.M.McKinley and F.C.Jones, Ap.J. 256,238 (1982).

[23] F.Winterberg, Phys.Rev. A19,1356 (1979).

[24] A.P.Mills,Jr., P.M.Platzman and B.L.Brown, Phys.Rev.Lett. 41, 1076 (1978); A.P.Mills,Jr., Appl.Phys.Lett. 35,427 (1979); ibid. 37,667 (1980).

[25] C.A.Murray and A.P.Mills,Jr., Solid State Commun. 34,789 (1980).

[26] J.M.Dale, L.D.Hulet and S.Pendyala, Surface and Interface Analysis 2,199 (1980).

[27] R.J.Wilson and A.P.Mills,Jr., Phys.Rev. B27,3949 (1983).

[28] P.J.Shultz, K.G.Lynn, W.Frieze and A.Vehanen, Phys.Rev. B27, 6626 (1983).

[29] S.Pendyala, P.W.Zitewitz, J.W.McGowan and P.H.R.Orth, Phys.Lett. 43A, 298 (1973).

[30] P.G.Coleman, T.C.Griffith and G.R.Heyland, Proc.Roy.Soc. London A331, 561 (1973).

[31] A.Vehanen, K.G.Lynn, P.J.Shultz and M.Eldrup to be published.

[32] P.B.Schwinberg, R.S.VanDyck, Jr. and H.G.Dehmelt, Phys.Rev. Lett. 47,1679 (1981) and refs. therein.

[33] J.H.Malmberg and J.S.deGrassie, Phys. Rev.Lett. 35,577 (1975)

[34] J.H.Malmberg and T.M.O'Neil, Phys.Rev.Lett. 39,1333 (1977).

[35] T.M.O'Neil, Phys.Fluids 23,725 (1980).

[36] W.M.Fairbank, F.C.Witteborn, J.M.J.Madey and J.M.Lockhart, Experimental Gravitation, Proc.Int.Sch.Phys. "Enrico Fermi" Course LVI, B. Bertotti, ed. (Academic Press, New York, 1974) p.310.

[37] K.G.Lynn, D.N.Lowy and I.K. MacKenzie, J.Phys. C: Solid St. Phys. 13, 919 (1980).

[38] It is unfortunate that one apparently cannot use the resistive tube damping idea of Ref. 36 to remove the longitudinal kinetic energy of the particles. Neglect of the skin depth in the calculation causes the resistive damping rate to be greatly over estimated. However, once the beam has internally thermalized it can be slowed to very low energies by a retarding field.

[39] K.G.Lynn, private communication.

SURFACE STUDIES WITH SLOW POSITRON BEAMS

R. M. Nieminen

Department of Physics
University of Jyväskylä, 40100 Jyväskylä, FINLAND

INTRODUCTION

Slow-positron physics is an exciting and rapidly advancing field. The continuing progress in the development of intense mono-chromatic beams of low-energy positrons has made it possible to per-form a number of landmark experiments, where the interaction of the positron with solid surfaces plays a central role. These experi-ments either deal with fundamental atomic physics (positronium spectroscopy) or focus on the electronic and atomic properties of the surface region, using positrons as a probe. In the former cat-egory, the surface is involved just as an efficient source of posi-tronium-like atoms. On the other hand, in the second category of experiments the surface is the main object of study, and has to be prepared and maintained under carefully monitored conditions. The challenge is then to understand the various aspects of the positron-surface interaction and to correlate the observations with micro-scopic surface information. I shall concentrate on discussing the positron surface physics from a theoretical viewpoint, with empha-sis on the new developments and future prospects. For background material and more detailed discussion, the reader is referred to a number of recent reviews[1-3].

POSITRON-SURFACE INTERACTION

The various physical processes, which a positron beam hitting a condensed matter surface may lead to, are schematically depicted in Fig. 1. Firstly, some of the incident positrons will reflect elastically from a few outermost atomic layers and form coherently diffracted beams in the case of a single crystal, in analogy with low-energy electron diffraction (LEED). The intensity of low-energy

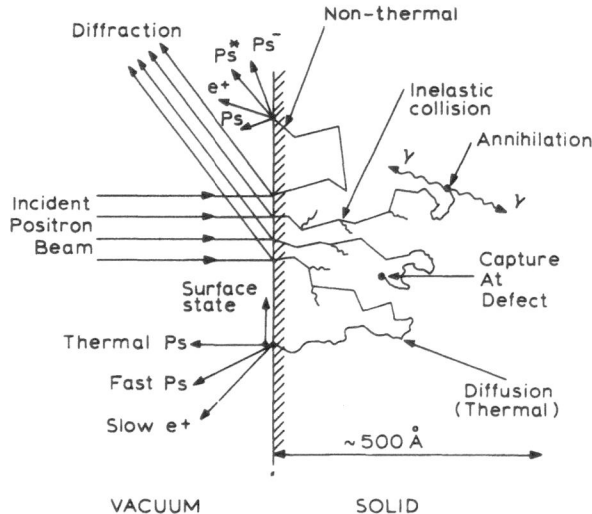

Fig. 1.   Schematic view of the positron surface processes.

positron diffraction[4,5] (LEPD) is highest (up to the order of a per
cent of the incident flux) in the energy range of a few hundred eV,
where the wavelength matches with typical lattice dimensions, and
where the inelastic mean free path is only a few Ångström.   The majority of the incident positrons will penetrate into the material
where they lose energy in various inelastic processes.   The time
scale for the slowing down is usually short compared with the total
positron lifetime.   However, at low temperatures and/or with low incident energies the thermalization may not always be completed, since
the particles can escape through the surface prior to thermalization[6].
The escape probability may also be a function of velocity and temperature, which leads to anomalies in the apparent positron equilibrium distribution[7]("positron fever").

    The implanted, thermalized positrons diffuse in the solid, where
they may be annihilated by an electron, be captured into a vacancy-
type defect or reach back to the entrance surface.

    For thermalized positrons diffusing near the surface several
fates are possible.   If they have a negative affinity (work function
ϕ_+ <0) for the solid, they may be spontaneously ejected into the vacuum[8].   This mechanism (denoted by "slow e$^+$" in Fig. 1.) is of course
central to making the slow beam in the first place.   Another important process is spontaneous Ps formation and emission[9] ("fast Ps" in
Fig. 1.), which is energetically allowed, since the Ps binding energy
E_B = 6.8 eV is larger than typical electron work functions ϕ_-.   Ps

emission is the only spontaneous emission route available if the
positron work function happens to be positive.

A third process is the positron capture into a potential well
induced by its own image interaction with the semi-infinite solid
just outside the surface.[10] The "surface state" (Fig. 1.) may subsequently be desorbed either thermally or by some other means of excitation. In thermal desorption[10,11] it is important to notice that
the emerging species is Ps, since $\phi_- < E_B$ usually. This is denoted
"Thermal Ps" in Fig. 1.

At low incident energies and/or temperatures (or in thin films)
a substantial fraction of the inelastically scattered positrons cross
the surface while unthermalized. This leads to spontaneous fast
positron and Ps emission, and (with a much smaller probability) the
emission of the positronium negative ion[12] (Ps$^-$), or of positronium
in one of its excited states (Ps*). These latter mechanisms provide
an example of the various possibilities which slow positron beams
offer to fundamental atomic physics investigations.

Another family of interesting phenomena are the inelastic processes that positrons lead to right at the surface. This includes
the secondary electron emission when the beam first strikes the surface. The secondaries can be used as a "start" signal in positron
lifetime experiments. Also the outgoing positron scatters inelastically[13] when spontaneously ejected from a negative-workfunction
solid. Based on this, reemitted-positron energy-loss spectroscopy[14]
(REPELS) has recently been demonstrated to be a probe for vibrational
excitations of adsorbed molecules on metal surfaces, complementary
to high-resolution electron energy loss measurements.

The slow-positron beams originating from β-active sources can
retain a high degree of spin-polarization,[15] as high as 60 %. This
makes it possible to investigate magnetic phenomena at surfaces. By
measuring the asymmetry in formation of the triplet spin state of
Ps, when either the surface magnetization or the positron-beam po-
larixation is reversed, the spin-polarization of the electron picked
up to form Ps can be determined. The first experiments[16] of this
kind show considerable promise for future work.

Low-energy Positron Diffraction

LEED has been used extensively in studies of atomic structures
on surfaces. While the information on lateral periodicity is easily
obtained and interpreted, the determination of perpendicular dis-
placements and overlayer structures is much more difficult. This
is due to the strong scattering of electrons in the region near the
ionic cores, where exchange and correlation effects are substantial;
the solution of the multiple scattering problem and its interpretation may be quite sensitive to even minute changes in the potential

construction. Therefore low-energy positron diffraction offers a
complementary tool for structure determinations, as has been demonstrated by the pioneering work at Brandeis[4].

The potential construction is less ambiguous for positrons than
for electrons. No exchange term arises in the positron-electron
scattering. The positrons are also strongly <u>repelled</u> electrostatic-
ally from the ionic cores, and mainly move in the interstitial regions where the electron density is slowly varying. Thus a local-
density approximation for the correlation potential[17] in LEPD is
quite adequate.

Another useful observation is that the relative strength of scattering of positrons and electrons off surface atoms depends on the
atom as well as on the incident energy. For heavy atoms the forward
scattering cross section is typically much higher for electrons than
for positrons; for light atoms the situation is reversed. Therefore
it could be possible to exploit these differences in systems which
consist of more than one kind of atom, e.g. adsorbate systems.

A point worth noting is that the inelastic mean free path is
predicted[18] to be shorter for positrons than for electrons, the difference ranging from around 30 % at 50 eV to 5 % at 100 eV. This
causes a substantial reduction in the contribution of subsurface
layers to the total diffracted intensity, and accordingly brings
about a reduction in the computing time required for convergence in
the multiple-scattering calculations[5]. The improved surface sensitivity of LEPD over LEED and the concomitant decrease in computing
requirements should be included in the comparison of the merits of
the two techniques.

Implantation and Diffusion

For a quantitative analysis of positron surface phenomena it is
important to understand the processes which govern the slowing down
and the subsequent diffusion stage.

At initial energies around a few keV, the positron loses its
energy quickly in ionizing collisions with the atoms and in inelastic scattering with the conduction electrons. The energy loss rate
at high energies can adequately be estimated from Bethe-type formulae.
In metallic targets, the inelastic scattering rate off conduction
electrons can be obtained from positron self-energy calculations[6,17,18]
for an electron gas. Typically one finds an energy loss rate $dE/dt \sim$
10^{17} eV sec^{-1}, with a maximum (minimum mean free path) around energy
of 100 eV. Consequently the time it takes to slow down from an in-

itial energy E of a few keV to an energy E_1 of a few tens eV is of
the order of 10^{-14} sec, and the total traversed path length $\Delta s \sim$
500Å $\cdot [E(keV)]^2$. Since the positron is also changing its momentum
(both in inelastic and elastic collisions), this is an upper bound
on how far it moves in the solid. In the transition energy region
between E_1 and near-thermal energies E_{th} (a few ten meV) the energy
loss rate is a rather complicated combination of electron and phonon
scattering[6], and the later stages of thermalization need careful calculations[6]. However, an order-of-magnitude estimate yields the ther-
malization time between E_1 and E_{th} as $\Delta t \sim 5 \cdot 10^{-12}[E_{th}(eV)]^{-3/2}$ sec,
and the traversed path length at this stage $\Delta s \sim 3Å/E_{th}(eV)$. We can
thus conclude that (i) the positron slows down very quickly to energies of a few ten eV; (ii) its range is nearly saturated by that
time; and (iii) the later stages of thermalization cause uncertainties at the level of less than 100Å in the stopping profile. In the
following we denote this profile (the distribution of trajectory endpoints as a function of the distance z from the surface) as P(z,E).

Valkealahti and Nieminen[19] have recently performed detailed calculations of the slowing down stage. They include both inelastic
and elastic processes in a Monte Carlo simulation of the collision
sequence. The scattering cross sections are evaluated in detail.
For elastic scattering off the lattice atoms, the cross sections are
calculated using a partial wave analysis based on self-consistent
atomic potentials. The inelastic scatterings with both bound electrons (ionizations and excitations) and conduction electrons (single-
particle and plasmon losses) are treated by using Gryzinski's expressions[20] for the relevant excitation functions. Useful information is obtained not only of the stopping profile P(z,E) but also
of the backscattering and thin film transmission yields, their energy and angular distributions, of the slowing down times etc.

Fig. 2 shows a projection of the endpoints of the simulated trajectories on the (y,z)-plane for 2-keV positrons impinging on Al.
Fig. 3 shows the analytic fits to the simulated implantation profiles
in a number of materials. A fairly good analytic approximation to
the profile is

$$P(z,E) = \frac{m}{z_0} (\frac{z}{z_0})^{m-1} e^{-(\frac{z}{z_0})^m} \tag{1}$$

where m and z_0 are functions of E and the target material. Typically
m $\sim$ 1.9 and depends weakly on energy and target material. If measured in units of mass per unit area, z_0 also is weakly dependent on
material. It increases with energy proportional to a power n. Consequently a good approximation to the mean penetration depth is

$$\bar{z} \cong 0.88 z_0 = \alpha E^n \tag{2}$$

where $\alpha \sim 4$ μg/cm^2 and n $\sim$ 1.6. More details can be found in Ref.19.

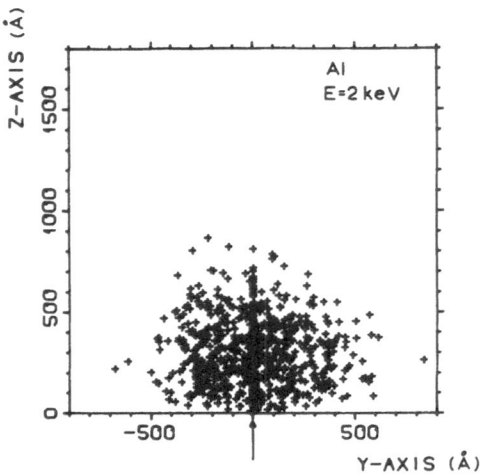

Fig. 2. The distribution of the trajectory endpoints, projected
         onto the y-z plane, for 2-keV positrons impinging on
         Al. The arrow denotes the beam position. From Ref.19.

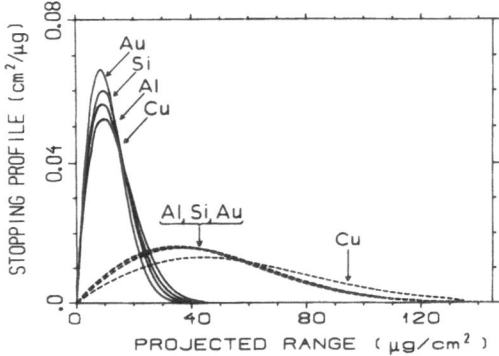

Fig. 3. Monte Carlo stopping profiles fitted to Eq. (1) for
         2.1-keV (full line) and 5.0 -keV (dotted line) positron
         in semi-infinite targets. From Ref. 19.

The simulation results can be compared to experimental transmission[21] and backscattering studies. An important lesson to be
learned from these comparisons is that the stopping profiles deduced
from transmission measurements carried out for varying film thicknesses can be quite different from those in semi-infinite targets.
The main reason for this is fairly strong dependence of the backscattering probability on film thickness, especially in very thin

films. A general conclusion for the implantation problem is that experiment and theory are now converging on each other, and that reliable values for the stopping profiles are becoming available from the simulations.

At completion of thermalization the positron diffuses in thermal equilibrium, scattering dominantly from phonons and impurities. To a reasonable approximation this motion is equivalent to an isotropic random walk, and it makes sense to describe in terms of a diffusion equation

$$D_+ \nabla^2 n(\bar{r},t) - \lambda(\bar{r})n(\bar{r},t) = \frac{\partial n(\bar{r},t)}{\partial t}, \tag{3}$$

where $n(\bar{r},t)$ is the positron density, D_+ is the diffusion constant and λ the positron depletion rate. It is the sum of the annihilation rate constant λ_a and the trapping rate γ to crystalline defects; the latter can be spatially varying due to a defect concentration gradient:

$$\lambda(\bar{r}) = \lambda_a + \gamma(\bar{r}) . \tag{4}$$

A most useful aspect of the slow positron technique is exactly its capacity to yield information on the defect profile near surfaces and interfaces.

One should note that the diffusion approximation for the positron density is reasonable only over distances greater than the mean free path between scatterings. For phonon-limited motion this is of the order of 10 Å at room temperatures. A good approximation for the diffusion constant due to positron-phonon scattering is[22]

$$D_+ \cong \frac{\sqrt{3}\pi}{3} \frac{e\hbar^4 c^2 \rho}{\Xi_d^2 m^{*5/2}} \frac{1}{(k_B T)^{\frac{1}{2}}} , \tag{5}$$

where c is the longitudinal sound velocity, ρ the material density, and Ξ_d and m^* are the positron deformation potential and effective mass, respectively. Typically D_+ is of the order of 1 cm^2 s^{-1} at room temperature. Since a free positron lifetime τ is a few hundred psec, the diffusion length prior to annihilation $L = \sqrt{6D_+\tau}$ is a few thousand Å. At low temperatures impurity scattering starts eventually to dominate, and the diffusion constant starts to decrease as proportional to $T^{\frac{1}{2}}$. The temperature T_c for crossover from phonon-limited to impurity-limited motion depends on the ratio $(\Delta V/\Xi_d)^2$, where ΔV is the impurity scattering potential averaged over the unit cell. T_c is typically a few $^\circ$K/at. % impurities.

In most experiments with slow positron beams a key quantity of interest is the re-emission yield of positrons or Ps atoms at the entrance surface. Within the diffusion approximation, it is straight-forward to calculate the yield

$$F = \nu \int_0^\infty dt \ n(z=0,t) \tag{6}$$

by solving the diffusion equation. Here ν is the total rate constant representing all processes capable of removing a positron from the solid. We assume that the planar surface is at $z = 0$ and that the positron thermalizes instantly in entering the solid. Thus the in-itial condition to Eq. (3) is

$$n(z,t=0) = P(z,E) \ .$$

The relevant boundary condition is associated with the surface, and reads

$$D_+ \left. \frac{\partial n(z,t)}{\partial z} \right|_{z=0} = \nu \ n(z,t) \big|_{z=0} \quad , \tag{7}$$

i.e. the re-emission current is the density at the surface times the rate constant ν. The yield is[6,23]

$$F = \nu \left[\frac{1}{(\lambda D_+)^{\frac{1}{2}}} - \frac{\nu}{\nu(\lambda D_+)^{\frac{1}{2}} + \lambda D_+} \right] I(E) \ ,$$

where

$$I(E) = \int_0^\infty dz \ P(z,E)e^{-sz} \quad \text{and} \quad s = \left(\frac{\lambda}{D_+}\right)^{\frac{1}{2}} \tag{9}$$

For a perfectly absorbing surface, $\nu \to \infty$ and $n(0,t) \to 0$, and $F = I$. By inserting the analytic fits to the implantation profiles in Eq. (9), one obtains the re-emission parameter I just as a function of $s \cdot z_0$. This is displayed in Fig. 4 for the case m = 1.9, which re-presents a large class of materials. Since the energy and material dependence of z_0 is known, Fig. 4 should be very useful in analysing re-emission data. It is also straightforward (but somewhat lengthy) to generalize the diffusion results to a case where there is a spa-tially varying depletion rate (a defect concentration gradient).

Slow Positron Emission

In direct analogy with its electron counterpart, the positron work function can be divided into a surface and bulk contribution[24]

$$\phi_+ = -D - \mu_+ \tag{10}$$

where D is the electrostatic surface dipole for electrons. Note that D and μ_+ in Eq. (10) have to have the same reference value;

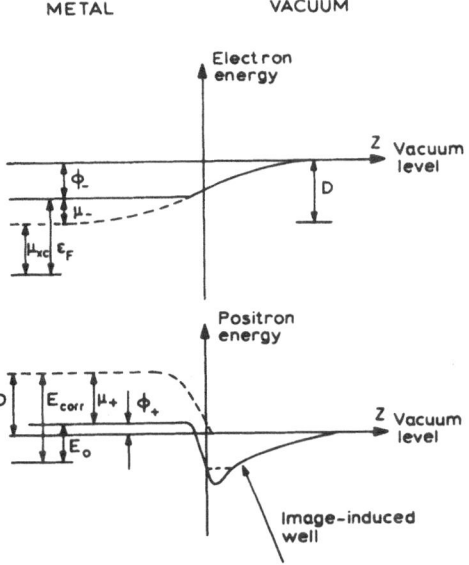

Fig. 4. Positron re-emission parameter I vs. the product of
         the diffusion parameter s and the penetration z_0.
         From Ref. 19.

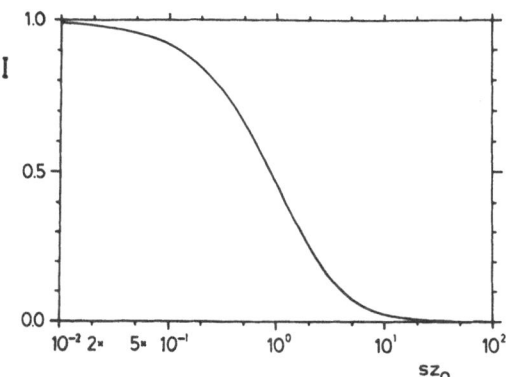

Fig. 5. Schematic view of electron and positron ground state
         energetics at a metal surface. D is electrostatic
         surface dipole, ε_F the electron Fermi energy, μ_{xc} the
         electron exchange-correlation potential and $\mu_- = \mu_{xc} +$
         ε_F the electron chemical potential. For positrons,
         $\mu_+ = E_{corr} + E_0$, where E_{corr} is the (negative) elec-
         tron-positron correlation energy and E_0 the zero-point
         energy. ϕ_- and ϕ_+ are the electron and positron work
         functions, respectively.

the commonly adopted one is the "crystal zero", i.e. the mean electrostatic potential at the edge of a bulk Wigner-Seitz cell. In
this case D > 0 for electrons (see Fig. 5 for a schematic picture).
The chemical potential μ_+ is equal to the energy of the lowest positron state in the solid, measured from the crystal zero. It is obtained as the eigenvalue of the positron Schrödinger equation, and
is customarily divided into a zero-point kinetic energy contribution
due to electrostatic repulsion from the cores, and a (negative) correlation energy contribution due to electron-positron interaction.

Experimentally one can determine a negative ϕ_+ by measuring the
energy distribution of spontaneously ejected positrons using an electrostatic analyzer. The work function corresponds to the elastically
emitted positrons. Theory and experiment for clean surfaces are
generally in good agreement[1,2] which gives credence for the basic
concepts and warrants further studies for more complicated (e.g. ad-
sorbate-covered) surfaces.

Evidence has been accruing that the emerging slow positron flux
can have a substantial low-energy tail in the perpendicular kinetic
energy distribution[13,25] One explanation could be surface roughness
or faceting, which makes the positrons escape with a wide angular
spread. A more likely alternative is, however, inelastic scattering
off electrons and surface excitations when the positron leaves the
solid. The mechanism is quite reminiscent of the dissipation processes which govern the positron trapping rate to defects. The inelastic effects have been considered theoretically by Pendry[26] and
by Jester et al.[27] The latter calculate the electronic friction due
to electron-hole pair excitations and find, in accord with experiment, that the inelastic tail increases when the positron work function becomes more negative.

An interesting recent development[14] is the observation of vibrational excitations of adsorbed CO on Ni(100) in the energy-loss
spectrum of re-emitted positrons. The outlook for using the technique in surface vibrational spectroscopy is good, and calls for
theoretical studies of the positron-surface molecule scattering in
both the small-angle (dipolar) and large-angle (impact) scattering
regimes.

Spontaneous Ps Emission

Positronium formation via electron pickup at the surface is energetically favorable, since the Ps binding energy E_B = 6.8 eV is
larger than typical electron work functions ϕ_-. With a smaller prob-

ability, also the positronium negative ion Ps⁻ or positronium in one
of its excited states may form. The Ps formation occurs diabatically:
its probability decreases when the positron work function becomes
more negative and thus the ejection velocity higher. This is due
to increasing diabaticity: the surface electrons have less time to
hop on to form the Ps. The theory of the process is reminiscent of
those of ion emission and neutralization, and the following expression can be derived[6] for the pickup probability P as a function of ϕ_+:

$$P = \frac{2}{\pi} \arctan \left(\exp[(\phi_0/\phi_+)^{\frac{1}{2}}] \right) \cong 1 - \frac{2}{\pi} \exp \left[-(\phi_0/\phi_+)^{\frac{1}{2}} \right] , \qquad (11)$$

where ϕ_0 is a constant characteristic of the material. Experiments
agree with Eq. (11); for Cu(111) covered with a partial layer of sul-
phur measurements[28] give $\phi_0 \cong 0.27$ eV, which is similar in size to
the theoretical estimates.

In cases where the positronium work function happens to be positive, Ps formation is the only spontaneous emission route (P = 1).
The only competing process is then the capture into a positron surface state, which will be discussed below.

An interesting application of the polarization of the incident
positron beam is the recent measurement by Gidley et al.[16] Since the
polarization is retained during slowing down and thermalization, one
can measure the <u>electron</u> spin-polarization at the surface. This is
because the formation probability of <u>triplet</u> Ps, which can easily
be determined, depends on the product of electron and positron spin-
polarizations. The asymmetry produced then by reversing the positron
polarization yields the spin polarization of the captured electron.

<u>Positron Surface States</u>

One particularly interesting facet of the positron-surface
interaction is the image-potential-induced surface state, which has
been the subject of extensive recent research, both experimental and
theoretical. As a charged particle a positron outside the surface
feels an attractive image interaction with the semi-infinite solid,
as is schematically shown in Fig. 5. This image attraction may be
strong enough for the positrons actually to become localized in a
surface state, as was first proposed by Hodges and Stott.[10] The ensuing quantum-mechanical problem, which is similar to the surface
polaron, is a challenging one since it embodies the full dynamic
and non-local subtleties of positron-electron correlation at the
surface.[10,29-31]

Since the experimental confirmation of positron surface states some years ago, high-quality experimental data about their properties has been accruing for well-defined surfaces. This facilitates comparison with theoretical calculations, which have become rather sophisticated and include now also the true geometric structure of the surface in question. The experimental observation is that the fraction of Ps produced at the surface rises drastically with temperature, following an Arrhenius-type curve. This can be explained in terms of thermal desorption of Ps, where the positron is taken from the surface state and combined with an electron from near the Fermi level to form Ps in vacuo. Classical thermodynamic desorption theory[32] then yields the activation energy for the process

$$E_a = E_b + \phi_- - E_B \quad , \tag{12}$$

where E_b is the surface state binding energy with respect to vacuum, and the desorption rate

$$\Gamma = Ae^{-E_a/k_B T} \tag{13}$$

The prefactor A depends on both the surface state characteristics and the reflection coefficient of the surface to Ps outside.[32] In order for the desorption process to become visible, the rate constant Γ has to be comparable to the inverse lifetime against annihilation in the surface state. Theoretical estimates[10,33] are available for the latter: the lifetime in the surface state is predicted to be substantially longer than the bulk lifetime. It will soon be possible to test these predictions as experimental facilities equipped with timing capabilities are becoming available.

An example of the power of the positron surface state spectroscopy is the study of oxidation of Al surfaces.[34] Oxygen at monolayer coverages is known to destroy Ps thermal desorption signal on all three principal surfaces, i.e. the surface state is unstable against spontaneous Ps emission, already at the lowest measuring temperature. Furthermore, the growth of a thick amorphous oxide layer produces defects at the metal-oxide interface. The annealing of these can be monitored by the beam technique.

Recently, a versatile calculational procedure[35] has been developed for obtaining quantitatively accurate positron states and their annihilation characteristics in various types of defects. By using three-dimensional numerical relaxation techniques, the program embodies the full geometry of the defect involved. As applied for image-trapped surface states, these calculations,[33] which reproduce well the observed activation energies on different clean surfaces, show that the disappearance of the surface state gives information about the aluminium-oxygen layer separation. On Al (100), the O

layer is most likely within the outermost Al plane. For Al(111) and
Al(110), the calculated minimum layer separation is 0.6 Å and 0.3 Å,
respectively.

The theoretical work also predicts that trapping by monovacancies
will be very weak on surfaces. Lateral localization is though very
likely to be caused by more extended surface defects. Furthermore,
the surface diffusion, as limited by phonons, is predicted to be tem-
perature-independent and faster than in bulk. Impurity scattering
on surfaces is relatively more important than in bulk. These investigations show that the sensitivity of Ps desorption studies is high
enough so that, in conjunction with a detailed calculational procedure, the observed parameter values can be correlated with atomis-
tic surface information. Lifetime studies will increase the information considerably.

Levine and Sander[36] have proposed an alternative model for the
desorption of surface bound positrons. They view the adsorbed
species as Ps, not positron. The emission process is then linked
with an excitation of the positron to its first excited surface state
and subsequent tunneling of the electron-positron pair. The model
requires that the excited state has to be rather stable and have a
sharp energy (the activation energy is in their model exactly the
excitation energy). Since little is known about the excited states,
the quantitative assessment of the model is difficult. On the other
hand, the classical desorption model seems to get strong support
from the time-of-flight measurements[37] for the velocity distribution
of thermally desorbed Ps.

CONCLUSION

In this paper I have described the basic physical concepts of
slow positron beams interacting with solid surfaces. The intensive
research during the last few years has produced many exciting developments. Undoubtedly this trend will continue in the near future.
Progress in experimental techniques will bring challenges for quantitatively accurate theoretical work. Applications to systems more
complicated than clean single crystal surfaces are only in their
beginning. Important recent work on glassy metals[38] and bimetallic
interfaces[39] are good examples of the possibilities. There is probably a lot to be done in the fields of thin film characterization
and metal-semiconductor heterostructures, just to give two examples.
The slow positron technique has the potential to become a rather
unique condensed matter tool, especially as regards structural and
electronic properties in the surface and near-surface region.

This work has been in part supported by the Academy of Finland.

References

1. See the articles by A. P. Mills, Jr., K. G. Lynn, and R. M. Nieminen, in "Positron Solid State Physics", W. Brandt and A. Dupasquier, eds., North-Holland, Amsterdam (1983).

2. A. P. Mills, Jr., in "Positron Annihilation", P. G. Coleman, S. C. Sharma, and L. M. Diana, eds., North-Holland, Amsterdam (1982).

3. K. G. Lynn, Scr. Met. 14, 9(1980); R. M. Nieminen, Phys. Scr., to be published.

4. I. J. Rosenberg, A. H. Weiss, and K. F. Canter, Phys. Rev. Lett. 44, 1139 (1980); A. H. Weiss, I. J. Rosenberg, K. F. Canter, C. B. Duke, and A. Paton, Phys. Rev.B 27, 867 (1983).

5. F. Jona, D. W. Jepsen, P. M. Marcus, I. J. Rosenberg, A. H. Weiss, and K. F. Canter, Solid State Commun. 36, 957 (1989); R. Feder, Solid State Commun. 34, 541 (1980); M. N. Read and D. N. Lowy, Surf. Sci. 107, L313 (1981).

6. R. M. Nieminen and J. Oliva, Phys. Rev.B 22, 2226 (1980).

7. W. Brandt and N. Arista, Phys. Rev. A 19, 2317 (1979); Phys. Rev. B 26, 4229 (1982).

8. A. P. Mills, Jr., P. M. Platzman, and B. L. Brown, Phys. Rev. Lett. 41, 1076 (1978).

9. A. P. Mills, Jr., Phys. Rev. Lett. 41, 1828 (1978).

10. C. H. Hodges and M. J. Stott, Solid State Commun. 12, 1153 (1973); R. M. Nieminen and C. H. Hodges, Phys. Rev. B 18, 2568 (1978).

11. K. G. Lynn, Phys. Rev. Lett. 43, 391 (1979); A. P. Mills, Jr., Solid State Commun. 31, 623 (1979); C. A. Murray and A. P. Mills, Jr., Solid State Commun. 34, 789 (1980).

12. A. P. Mills, Jr., Phys. Rev. Lett. 46, 717 (1981); ibid. 50, 671 (1983).

13. R. J. Wilson, in "Positron Annihilation", P. G. Coleman, S. C. Sharma, and L. M. Diana, eds., North-Holland, Amsterdam (1982); R. J. Wilson and A. P. Mills, Jr., Phys. Rev. B 27, 3949 (1983).

14. D. A. Fischer, K. G. Lynn, and W. E. Frieze, Phys. Rev. Lett. 50, 1149 (1983).

15. P. W. Zitzewitz, J. C. Van House, A. Rich, and D. W. Gidley, Phys. Rev. Lett. 43, 1281 (1979).

16. D. W. Gidley, A. R. Köynmen, and T. W. Capehart, Phys. Rev. Lett. 49, 1779 (1982).

17. J. Arponen and E. Pajanne, Ann. Phys. 121, 343 (1979); J. Phys. F. 9, 2359 (1979).

18. J. Oliva, Phys. Rev. B 21, 4909 (1980); G. C. Aers and J. B. Pendry, J. Phys. C 15, 3725 (1982).

19. S. Valkealahti and R. M. Nieminen, Appl. Phys. (in press), and to be published.

20. M. Gryzinski, Phys. Rev. A 138, 305 (1965).

21. A. P. Mills and R. J. Wilson, Phys. Rev. A 26, 490 (1982).
22. B. Bergersen, E. Pajanne, P. Kubica, M. J. Stott, and
    C. H. Hodges, Solid State Commun. 15, 1377 (1974).
23. A. P. Mills, Jr. and C. A. Murray, Appl. Phys. 21, 1
    (1980); K. G. Lynn and D. O. Welch, Phys. Rev. B 22,
    99 (1980).
24. C. H. Hodges and M. J. Stott, Phys. Rev. B 7, 73 (1973);
    R. M. Nieminen and C. H. Hodges, Solid State Commun.
    18, 1115 (1976); R. M. Nieminen and C. H. Hodges, J.
    Phys. F 6, 573 (1976); G. Fletcher, J. L. Fry, and P.
    C. Pattnaik, Phys. Rev. B 27, 3987 (1983).
25. R. J. Wilson, private communication.
26. J. B. Pendry, J. Phys. C 13, 1159 (1980).
27. K. Jester, D. Neilson, and R. M. Nieminen, unpublished.
28. A. P. Mills, Jr. and C. A. Murray, Bull. Am. Phys. Soc.
    25, 392 (1980).
29. R.M. Nieminen and M. Manninen, Solid State Commun. 15,
    403 (1974); N. Barberan and P. M. Echenique, Phys.
    Rev. B 19, 5431 (1979); J. E. Inglesfield and M. J.
    Stott, J. Phys. F 10, 253 (1980).
30. M. Babiker and D. R. Tilley, Solid State Commun. 39, 961
    (1981); M. Babiker, Physica 103B, 289 (1981); G. Barton
    and M. Babiker, J. Phys. C 14, 4951 (1981).
31. G. Barton, J. Phys. C 14, 3975 (1981); J. Phys. C 15,
    4727 (1982).
32. S. Chu, A. P. Mills, Jr., and C. A. Murray, Phys. Rev.
    B 23, 2060 (1981); M. Manninen and R. M. Nieminen,
    Appl. Phys. A 26, 93 (1981).
33. R. M. Nieminen and M. J. Puska, Phys. Rev. Lett. 50, 281
    (1983).
34. K. G. Lynn, Phys. Rev. Lett. 44, 1330 (1980); K. G. Lynn
    and H. Lutz, Phys. Rev. B 22, 4143 (1980).
35. M. J. Puska and R. M. Nieminen, J. Phys. F 13, 333 (1983).
36. R. Y. Levine and L. M. Sander, Solid State Commun. 42,
    5 (1982).
37. A. P. Mills, Jr. and L. Pfeiffer, Phys. Rev. Lett. 43,
    1961 (1979).
38. A. Vehanen, K. G. Lynn, P. J. Schultz, E. Cartier, H. J.
    Güntherodt, and D. M. Parkin, to be published.
39. P. J. Schultz, K. G. Lynn, W. E. Frieze, and A. Vehanen,
    to be published.

INTENSE POSITRON BEAMS: LINACS*

R. H. Howell and R. A. Alvarez

Lawrence Livermore National Laboratory
Livermore, CA 94550

K. A. Woodle, S. Dhawan, P. O. Egan,
V. W. Hughes, and M. W. Ritter

Gibbs Laboratory, Yale University
New Haven, CT 06520

SUMMARY

Beams of monoenergetic positrons with energies of a few eV to many keV have been used in experiments in atomic physics, solid state physics and materials science. The production of positron beams from a new source, an electron linac, is described.

Intense, pulsed beams of low-energy positrons have been produced by a high-energy beam from an electron linac. The production efficiency, moderator geometry, beam spot size and other positron beam parameters have been determined for electrons with energies from 60 to 120 MeV. Low-energy positron beams produced with a high-energy electron linac can be of much higher intensity than those beams currently derived from radioactive sources. These higher intensity beams will make possible positron experiments previously infeasible.

---

*Work performed under the auspices of the U.S. Department of Energy by the Lawrence Livermore National Laboratory under contract number W-7405-ENG-48.

INTRODUCTION

Since the introduction of experiments using beams of monoenergetic positrons there has been a continuing effort to increase
the intensity of these positron sources for both pulsed and
continuous beams. These efforts have centered mainly around the
improvement in the conversion of energetic positrons from radioactive sources to mono-energetic, low-energy positrons by improving
the moderator material characteristics. This effort has led to
higher beam intensity through improved moderator efficiency and a
clearer understanding of the production of low energy positrons by
a process of diffusion to the moderator surface and expulsion by
the negative positron work function. The efficiency of the
moderation process has been improved to such an extent that
theoretical limits have been approched. Any further large
increases in positron beam intensity must come from the use of
more intense sources.

To provide a more intense source of positrons through radioactive decay requires the use of sources with short half lives.
The limit of the positron intensity from the usual sources of
positrons, ^{22}Na or ^{58}Co, results from the inability of the
positrons from the back of the source to penetrate the source
material and escape the source configuration. The overburden of
material is lower in sources that have short half lives, ^{11}C or
^{64}Cu but these sources must be continually renewed by
irradiation with an accelerator[1] or reactor.[2]

Intense sources of high energy positrons can be obtained by
pair production in the bremsstrahlung field that results from
hitting a radiator-converter with the electron beam. Positrons
are produced in electron-positron showers resulting from repeated
cycles of pair-production and bremsstrahlung radiation. The positrons can be moderated and emitted as low energy positrons in much
the same way as in the radioactive source systems. Low energy
positron beams produced by this technique have been used at other
laboratories, but the beams were of low intensity.[3,4] More
recently work at Livermore[5] has demonstrated that intense beams
of low energy positrons can be derived from linac electron beams.
Experiments performed at Mainz[6] have also been successful in
producing intense positron beams.

In the Livermore experiments the important parameters of the
production of low energy positrons from high energy electrons have
been identified and operating ranges established: The geometry of
the electron radiator-convertor and positron moderator was studied,
and the optimum radiator-convertor thickness was determined for
both tungsten and tantalum convertors. The variation in positron
production with primary electron beam energy was measured. The

size of the positron emitting region on the moderator was
determined for several incident primary beam angles. Some of the
potential problems in routine production of intense beams were
identified.

It was found that the most important parameter affecting
low-energy positron production is the geometry of the radiator-
converter and moderator. The best results were obtained with the
closest coupling possible between these two elements. The optimum
thickness of the radiator-converter is the same number of radiation
lengths for the materials studied and the overall positron production was dominated by the power deposited in the radiator-converter
regardless of electron beam energy for energies above 60 MeV.

EXPERIMENTAL DETAILS

The initial Livermore experiments were performed with a
conventional bent solenoid slow-positron transport system, similar
to those described in Refs. 7 and 8. This apparatus was used to
measure the slow-positron production efficiency. The electron
beam passed through thin stainless steel windows to reach the
position of the electron-positron converter. The converter-
moderator assembly was withdrawn to tune the electron beam. A
bias between the moderator and an aperture accelerated the slow
positrons to energies between 10 and 100 eV. This transport
system had 50 percent transmission 1 cm off the central axis.

Initially, positrons were detected by a 1 cm diameter channel
electron multiplier (CEM) positioned on the solenoid axis. The
CEM was run in a single particle counting mode. Thus multiple
positrons in a beam pulse were indistinguishable from single
positrons. To measure the electron-beam-to-slow-positron conversion efficiency accurately, it was necessary to restrict the
electron beam current so that the positron counting rate in the
CEM was about 30 percent of the beam repetition rate of 1440 sec^{-1}.

A second system constructed at Yale allowed us to transport
the slow positrons out of the accelerator cave and away from most
of the beam-induced background. In this apparatus the positron
transport efficiency is unmeasured but the transmission of thermal
electrons is 86%. In the low background experimental area we used
both NaI detectors and micro-channel plate detectors to measure the
positron production rates with electron beam peak currents near the
maximum available from the linac. Observations of both the output
current trace of the micro-channel plate and the annihilation
radiation counting rate in the NaI detectors placed far from the
slow positron beam stop demonstrated that slow positron production
was proportional to the beam current for all beam levels available
at our linac.

A third positron transport system was constructed as a prototype of the positron production part of a permanently available
low energy positron beam line dedicated to materials science
experiments, Fig. 1. In the prototype system we could acurately
determine the size and shape of the positron emitting beam spot by
observing the optical output of a micro-channel plate. From the
electrical output of the micro-channel plate we measured positron
production efficiency and the distribution of the positron time of
flight.

A new feature of the prototype transport system was that the
radiator-convertor was outside the vacuum system and the moderator
inside. Vacuum was maintained by a thin aluminium window. The
transport system also has the capability of extracting and transporting the positrons using purely electrostatic elements or
magnetic elements or a combination of both. By adjusting the
fields in the prototype system it was possible to reduce the size
of the spot on the micro-channel plate to less than 1/4 of the size
of the positron emitting spot on the moderator.

In all transport systems the energy of the positrons was
measured by the time of flight. The agreement between the energy
measured by time of flight and the acceleration voltage uniquely
identified the slow positrons. There was also a prompt signal
produced by the intense bremsstrahlung flash from the primary linac
beam. The prompt signal served as a convenient time mark for the
time-of-flight determinations.

RESULTS

We found in all cases that close coupling between the electron
radiator-converter and positron moderator resulted in the highest
slow-positron yield. This is expected, since the positrons emerging
from the converter have a high angular divergence. The best
materials for radiator-converters have the combination of high
density, high atomic number, high melting point and low residual
radioactivity after use. For the radiator-converter we have used
either tungsten or tantalum at different times and found that the
positron yield is not sensitive to which of these materials were
used. Both of these materials have all of the desirable characteristic except that they have high levels of residual radioactivity.

For the positron moderator tungsten was chosen from among the
materials known to have high negative work functions for positrons
because of its high positron yield, ease of preparation, and
resistance to degradation in air. The higher atomic number of
tungsten gives it a higher stopping efficiency for energetic
positrons and a lower susceptability to radiation damage by
positrons or electrons that have only a few MeV of energy.

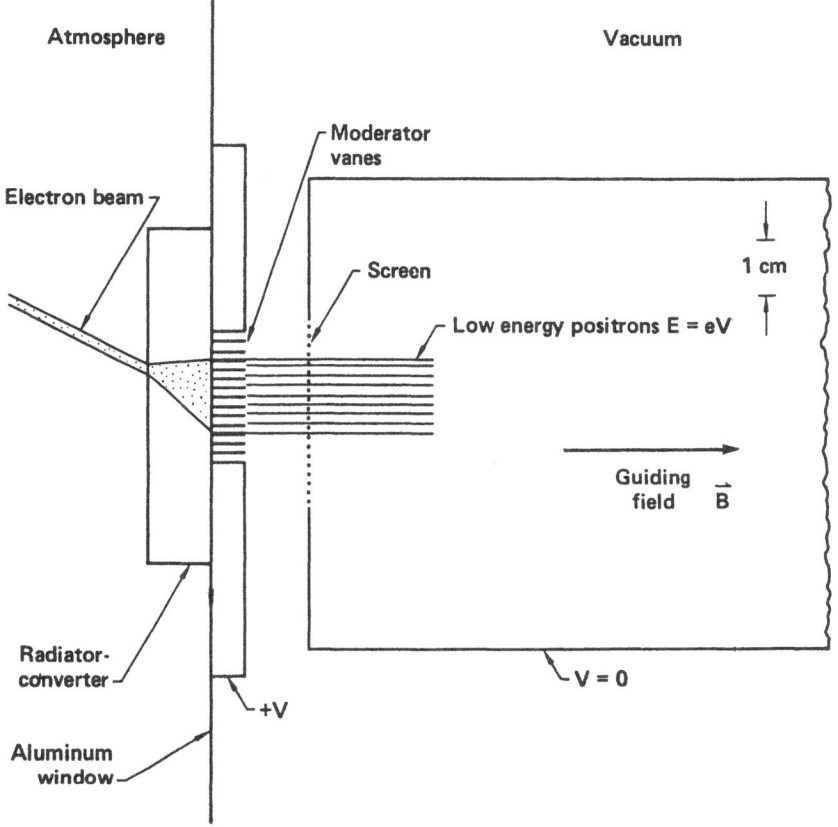

Fig. 1. Front end geometry of the prototype system designed to study design features for the permanent beam line being installed at Livermore. Problems of shielding from residual radiation and removal of heat are simplified by keeping the radiator-converter in air. Positrons were detected with a micro-channel plate 1.5 m from the moderator.

The moderator material was prepared as described in Ref. 9 and then stored and transferred in air to the experimental apparatus. Copper moderators also have a high positron yield, but must be prepared and stored in a high vacuum[10] and may not be as radiation resistant as tungsten.

The best geometry found for the moderator and transport consists of tungsten foils arranged as a series of vanes with the front edges facing the transport system and the surface of the vane parallel to the magnetic guiding field in the transport. Several other geometries for the moderator were tried including flat plates with the plate surface perpendicular to the guiding field and a series of wires arranged in a grid. Comparing these geometries, we found that the low energy positron yield was the highest with the vanes and was lower in the other geometries by about the same amount as the reduction in surface area of the moderators.

In Ref. 5 the positron yield is reported for different electron radiator-converter thicknesses; the best yield was found at three radiation lengths for electrons in tantalum. Similar measurements, shown in Fig. 1, were made with 100 MeV electrons and a tungsten radiator-convertor in the prototype transport system. The maximum low energy positron production is still found at a thickness corresponding to three radiation lengths in the material. Thus about 95 percent of the energy in the original electron beam has been converted into other forms of energy, including positrons.

The size of the positron emitting region of the moderator could be determined in the prototype transport system by observing the positron spot on the micro-channel plate. These measurements were done for a flat plate tungsten radiator-convertor, three radiation lengths thick, with the moderator vanes placed directly against the downstream side of the plate. The edge of the vanes was clearly visible in the positron beam spot so the vane spacing provided a length calibration. An electron beam about 2 mm in diameter was incident on the plate at either 40° or 15°. For both electron beam angles the size of the positron emitting region of the moderator was roughly 1 cm.

The highest efficiency to date was measured in the prototype transport system by reducing the electron beam intensity and counting single positron events in the micro-channel plate. The value is 2.0×10^{-6} slow positrons per incident electron. This efficiency is high enough that with the beam from the Lawrence Livermore National Laboratory S-band Linac, slow-positron beams of 0.5 namp average current are possible. With short pulse duration, beams with 8×10^{5} slow positrons per pulse can be produced in pulses shorter than 20 nsec repeated 1440 times a second. These beams are several orders of magnitude more intense than those from existing radioactive sources.

OTHER EXPERIMENTS

At the Mainz linac experiments were performed with an apparatus that was similar to that used in the Livermore experiments. In the Mainz experiments positron yield was measured by detection of annihilation gamma rays while either the energy or intensity of the electron beam was varied. The results from these measurements are generally consistent with the Livermore results. With 1 μA of 200 MeV electrons 6×10^5 positrons per pulse were transported to the sample area. The efficiency of the positron production at Mainz was 9×10^{-6} positrons/electron at 200 MeV electron energy. This is about 3 times better efficiency than the best Livermore results. Since the geometry of the radiator-moderator region is similar in the Mainz and Livermore experiments the higher efficiency in the Mainz experiment is probably the result of moderator baking that occured in the Mainz system but not in the Livermore system.

The Mainz system is being upgraded to accommadate higher electron beam intensities and pulses of more than $5 \; 10^7$ positrons in 1 μsec are expected.

DISCUSSION

Production of positron beams with an electron linac results in special properties and problems due to the properties of the linac electron beam. Since the electron beam in most linacs is pulsed the positron beam will be pulsed as well. The time width of the positron pulse will depend on the width of the electron beam pulse, the lifetime of the positrons in the moderator and the qualities of the positron transport. In a typical rf linac the electron beam has a time structure consisting of a train of micropulses each about 10^{-11} s in duration separated by the rf period. The pulse train may have any length from a nanosecond to several microseconds. Thus with proper bunching it is possible to have positron pulses that are short enough to use in short lifetime experiments. Bunching techniques can also be used to compensate for the energy dispersion of the slow positrons and so very good time definition or very good energy definition can be obtained in intense slow-positron beams.

With intense, pulsed slow-positron beams we can perform time-of-flight energy analysis of scattering from gasses and time-of-flight measurements of positrons and positronium scattered or diffracted in a wide variety of conditions. Also we can perform new materials-science measurements including two-dimensional angular correlation measurements, and with a short positron pulse, positron lifetimes. These and other materials measurements such as Doppler broadening can be made on samples during rapidly changing conditions to study transient effects.

There are two practical problems with producing high intensity
low energy positron beams with an electron linac: First is the need
to dissipate large amounts of power in the radiator-convertor and
also in the surronding apparatus. The LLNL linac produces 10 kw in
the mode that has a high rate of short pulses and nearly 40 kw in
the mode that has a low rate of longer pulses. In the production
of low energy positrons a large fraction of the linac power will be
deposited in the radiator-convertor in the form of heat which must
then be carried off. Second is the high levels of radioactivity
that will be present during and after the positron production.
Materials used in the construction of a low energy positron system
must be chosen with radiation hardness and residual activation
characteristics in mind. Organic materials degrade quickly when
near the linac beam, and steel and many other metals remain
radioactive for relatively long times after irradiation.

There are also problems in setting up experiments with linac
positrons. The intense bremsstrahlung produced during positron
production must be shielded for many experiments. Also when using
pulsed sources the performance is often limited by detection systems
that cannot accomodate multiple events in one beam pulse. In this
case the intensity of the source must be low enough that the
detector will not respond at every pulse. If this condition is not
met then the counting rate in the detector is just the source pulse
rate and the detector will respond preferentially to the early
events leading to a distortion of the data. In general the source
intensity must be limited so that about 30% of the source pulses
result in detector responses.

There is a solution to the intensity problem if the detector
is capable of responding to multiple events during a single beam
pulse. In that case the source strength is no longer limited by
the pulse rate and a complete measurement is possible from a single
beam pulse. At Livermore using a microchannel plate detector the
fraction of re-emitted positrons has been measured for a single
beam pulse. The fraction of positronium formed in a single beam
pulse may also be measured using fast plastic gamma-ray detectors.

At the Livermore linac there are two beam lines under
development for the transport of slow positrons. Both of these
will be based on the use of long solenoids for magnetic transport
of the low-energy positrons. One of these will be dedicated to
positronium atomic physics experiments in a collaboration with
groups at Yale and Stanford. The second transport will be for
general use in solid state physics and materials science
experiments. Eventually the materials science beam will have
several useful features including a bunching system for lifetime
experiments and sufficient energy range to cover both surface and
bulk measurements.

The atomic physics beam line has been constructed at Yale and is in final development at Livermore. The permanent materials science beam line has been designed and the basic system is under construction. The prototype of the positron production part of the transport has been used in tests and is now free for use in new experiments.

REFERENCES

1. W. E. Kaupilla, T. S. Stein and G. Jesion, Phys. Rev. Lett. 36, 580 (1976).
2. K. Lynn, see the article in this proceedings.
3. D. G. Costello, D. E. Groce, D. F. Herring, and J. Wm. McGowan, Phys. Rev. B 4, 1433 (1972); and L. S. Goodman and B. L. Donnally, unpublished.
4. M. Begemann, G. Gräff, H. Herminghaus, H. Kalinowsky, and R. Ley, Nucl. Inst. and Meth. 201, 287 (1982).
5. R. H. Howell, R. A. Alvarez and M. Stanek, Appl. Phys. Lett. 40, 751 (1982).
6. G. Gräff, R. Ley, A. Osipowicz, G. Werth, and J. Ahrens, to be published.
7. A. P. Mills, Appl. Phys. 22, 1 (1980).
8. K. G. Lynn and H. Lutz, Rev. Sci. Instrum. 51, 977 (1980).
9. J. M. Dale, L. D. Hulett, and S. Pendyala, Surface and Interface Analysis 2, 199 (1980).
10. A. P. Mills, Appl. Phys. Lett. 35, 427 (1979).

DISCLAIMER

INTENSE POSITRON BEAMS AND POSSIBLE EXPERIMENTS

K.G. Lynn and W.E. Frieze

Physics Department
Brookhaven National Laboratory
Upton, New York 11973 USA

INTRODUCTION

In this paper, we will survey some of the ideas that have
been proposed regarding the production of intense beams of low
energy positrons. Various facilities to produce beams of this
type are already under design or construction and other methods
beyond those in use have been previously discussed.(1) Moreover,
a variety of potential experiments utilizing intense positron
beams have been suggested. It is to be hoped that this paper can
serve as a useful summary of some of the current ideas, as well as
a stimulation for new ideas to be forthcoming at the workshop.

We begin with a general description of the development of
variable energy positron beams to date. A second section will
emphasize the intense positron beams currently being built.
Particular advantages and disadvantages will be discussed for the
methods to be used to produce positrons for these beams. Emphasis
will be placed on the beam under construction at the High Flux
Beam Reactor (HFBR) at Brookhaven. The third section of the paper
will briefly sketch some of the new condensed matter experiments
that will be possible in the future owing to the increased positron current. Some of these proposed experiments are feasible
immediately, while others must await implementation of certain
technological developments (such as brightness enhancement) (2)
for their completion.

PRESENT BEAMS

We make the distinction in this paper between laboratory
based beams, which are completely self-contained and utilize a

dedicated positron source, and facility based beams, which utilize an accelerator or reactor to provide an intense source of positrons. Slow positron experiments have thus far been done almost exclusively using laboratory based beams in which the number of particles varies from 10^4 to 5×10^6 per sec. This current of positrons is suitable for a large class of experiments as is demonstrated by the many measurements that have been done using them already (1,3). Some measurements however, do require higher positron rates and for these the new, intense beams are necessary.

Operationally, we will define intense positron beams as those that will supply more positrons than any presently existing laboratory beam using standard radioactive isotope sources (i.e. Co^{58}, B^{11} and Na^{22}). This definition is not strictly correct as sometimes facility based beams can have characteristics such as time bunching which can make them ideally suited for certain experiments even though the average current may be less than a standard laboratory beam.

The beams presently in use can be classified based on the source of the positrons, the type of moderator employed, and the transport system used, i.e. magnetic or electrostatic. In this paper we will not discuss general moderator development. Suffice it to say that low defect density polycrystalline (4) or single crystal (5) W samples appear to yield the largest moderation efficiencies as of this writing. Some recent experiments at Brookhaven seem to indicate that Pt(100) + CO (6) may also be competitive with single crystal W.

The type of transport system chosen will be dependent on the experiments planned. However, experience at Brookhaven has shown that systems using transport by a magnetic guiding field are much easier to construct and operate. (7) In this technique one uses an axial magnetic field to guide the positrons, caught in small cyclotron orbits, out of the high background source region and into a target chamber. Building a fully electrostatic beam is more complicated but allows more versatility; in these systems, electrostatic lenses of various types are used to focus and transport the extracted slow positrons. In the future, hybrid beams are likely to be developed where the moderated positrons will be magnetically transported into an electrostatic final section promising a compromise between ease and flexibility.

The source of positrons most commonly used is a commercially purified source, e.g. Co^{58}, which is plated onto a metal backing (8) or a salt of Na^{22} which is housed in a vacuum capsule or is incorporated into some medium such as glass or ceramic. These commercially purified sources are usually limited in total activity (<50mCi of positrons) and are rather expensive. Another source which has been utilized successfully is produced when protons(9) are used to irradiate a B^{10} target. Positrons are then emitted from C^{11} produced by the irradiation. A tabulation of all those positron sources which have a half-life greater than 12 hours is shown in Table 1, listed in order of the relative fraction of β^+ per disintegration.(10)

INTENSE POSITRON BEAMS

The production of intense positron beams ($\geq 5 \times 10^6$ positrons per sec) will probably require techniques not available in the laboratory. While many methods can be imagined which will produce large numbers of positrons, facility based beams using two different methods are currently being constructed. These two methods utilize either a linear electron accelerator (11,12,13) to produce electron-positron showers via pair production or a nuclear reactor to produce intense Cu^{64} sources by neutron capture.

In the first method, one generates an intense positron beam by moderating the positrons produced in the positron-electron showers created by the incident electron beam. (11) This method may prove to produce the largest total number of slow positrons of any method that has been suggested. Efficiencies between 10^{-6} and 10^{-4} positrons per incident electron have been reported under conditions of reduced primary beam intensity (11,12). Whether these efficiencies can be maintained as the beam intensity is increased remains to be demonstrated. In the case of increased electron beam intensity, considerable radiation damage and heating of the target occurs, both of which effects are known to reduce the efficiency of moderation (1). If these problems can be solved, however, a linac such as that at Livermore National Laboratory might conceivably yield as many as 10^{11} positrons per sec, a rate obtainable in reactor based beams only with sources in the 10 kCi range.

Beam rates as large as this may make linac based slow positron beams quite useful for certain experiments, but a number of disadvantages are also present. Perhaps the primary one is cost. Because a linac beam requires the full electron beam for maximum positron flux, simultaneous operation with other users is impossible. If one is familiar with the scheduling and other problems involved in accelerator use, the difficulty involved here is apparent (especially when coupled to experiments involving the chemistry or physics of surfaces where considerable sample preparation time is required while positrons are also available).

Another aspect of linac beams may be an advantage or a disadvantage, depending of the application. Because many of these machines are intrinsically time pulsed in nature, experiments which require a start signal when a positron arrives are simplified. However, other experiments will be much more difficult due to the very low duty cycle of these pulsed beams. Except for

Table 1:   Suitable Positron-Emitting Isotopes

Isotope	τ	β^+/dis	β^+	Production Reaction
Na^{22}	2.6y	0.89	0.54	$Mg^{24}(d,\alpha)$
Al^{26}	$7.4 \times 10^5 y$	0.85	1.17	$Mg^{24}(d,\gamma)$
Co^{55}	18.2h	0.60	1.50,1.03,0.53,0.26	$Fe^{12}(p,2n)$
V^{48}	16.2d	0.56	0.69	$Ti^{48}(p,n)$
Ni^{57}	36h	0.50	0.85,0.72,0.35	$Ni^{58}(p,pn)$
Sr^{83}	33h	0.50	1.15	$Sr^{84}(p,pn)$
Y^{86}	14.6h	0.50	1.80,1.19	$Sr^{86}(p,n)$
Br^{76}	17.2h	0.44	3.57,1.7,1.1,0.8,0.6	$Se^{76}(p,n)$
Nb^{90}	14.6h	0.40	1.51,0.66	$Zr^{90}(p,n)$
Mn^{52}	5.7d	0.35	0.58	$Cr^{52}(p,n)$
Ge^{69}	40h	0.33	1.22,0.61,0.22	$Ga^{69}(p,n)$
As^{71}	62h	0.30	0.81	$Ge^{72}(p,2n)$
As^{72}	26h	0.30	3.34,2.50,1.84,0.67,0.27	$Ge^{72}(p,n)$
I^{124}	4.5d	0.30	2.20,1.50,0.70	$Te^{124}(p,n)$
As^{74}	17.5d	0.29	1.53,0.92	$Ge^{74}(p,n)$
Zr^{89}	79h	0.25	0.91	$Y^{89}(p,n)$
Co^{56}	77d	0.20	0.44,1.50	$Fe^{56}(p,n)$
Cu^{64}	12.8h	0.19	0.65	$Cu^{63}(n,\gamma)$
Rb^{84}	33d	0.17	1.63,0.82	$Sr^{86}(d,\alpha)$
Co^{58}	71d	0.15	0.47	$Ni^{58}(n,p)$

the detection efficiency measurements, the presence of thousands
or even millions of positrons within a few tens or hundreds of ns
will cause serious problems with accidentals.

Certain other linac characteristics are probably of less significance, but deserve mentioning. Because of long lived isotopes
produced by the primary beam, access to the moderator will be very
limited, thus making changes, maintenance or repair more difficult. A related problem is that transport of the positron beam
out of the shielded target room is a fairly difficult and expensive operation. This will also cause time debunching due to the path
length required, and thus, some of the advantages of a pulsed
machine may be lost. Finally we note in passing that linac beams
are intrinsically unpolarized, a situation which may or may not be
a drawback.

We see then that linac based beams have a promising future
although a variety of problems need to be solved. Still, if the
primary problem, that of maintaining moderation efficiency as the
primary beam intensity is increased, can be solved then it is certain that these beams will be major contributors to the future of
slow positron physics.

A second method of producing copious amounts of positrons is
to produce the Cu^{64} isotope. This is done by thermal neutron capture by Cu^{63} in a nuclear reactor. A fortuitous aspect of this
method is that Cu^{64} has a short half-life (12.7h) and low background of gamma rays. This first characteristic reduces any problems due to long term contamination of an experimental apparatus.
Using the thermal neutron capture cross section for Cu^{63} of 4.5 b
and, for example, using the thermal neutron flux of $1 \times 10^{15} n/cm^2$-
sec available at the HFBR, we obtain an activity of 580 Ci per
gram of natural Cu after a 24 hr. irradiation. A 1 cm^2 source
with a thickness corresponding to the beta end point range in Cu
of 0.023 cm, would give an activity of 120 Ci, more than two
orders of magnitude larger in positron activity than the 0.2-0.5
Ci ^{58}Co sources used in many slow positron beams.

Taking an effective moderation efficiency of 1×10^{-3} and beam
transmission of 90% (and including a branching ratio of 19% for e^+
decay) we obtain a beam rate of 4×10^8 s^{-1} for a 1 cm^2 source,
around 100 times the rates measured in existing beams. As the
result of the 12.7 h half-life, the total number of slow positrons
available per non-enriched Cu source is 2.5×10^{13}.

This type of reactor based beam is ideally suited for continuous beam operation as no intrinsic time structure is present.
However, one is not limited to this type of experiment. For example, on the laboratory based electrostatic beam at Brookhaven we
have found that timing resolutions of the order of 600 psec full

width at half maximum (FWHM) can be obtained by simply using the
secondary electrons generated when positrons strike a target as a
start signal and the annihilation gamma ray as the stop signal.
(14,15) The secondary electrons were detected by a Channel Electron Multiplier Array and the annihilation photons by a BaF_2 scintillator crystal. This shows that one can do high resolution timing experiments on an unbunched beam.

The reactor method of production also allows for the capability of making highly spin polarized positron beams simply by using
a transmission type self-moderator thus increasing the ratio of
v/c in those positrons which become moderated. Such a moderator
could be produced by evaporating various amounts of non-radioac-
tive Cu as overlayer. This characteristic of producing a highly
polarized as well as an intense beam makes feasible many new
experiments. For example, one can perform an energy and angle
resolved measurement of emitted Ps (16,17) from ferromagnets.
This would provide more information on the nature of the electrons
involved in surface magnetism.(15)

A major advantage of a reactor based beam is that one can run
in a truly symbiotic mode and not have a measurable effect on
other reactor operations. This procedure also allows for a number
of simultaneous irradiations of different Cu^{63} samples to feed a
number of potential beams. Owing to the short half-life of Cu^{64}
and the intensity of the emitted radiation, all such beams must,
however, be very close to the irradiation ports of the reactor.
The cost per irradiation is relatively small in an operating reactor. One does need a large thermal flux to make the specific
activity of the Cu source adequate to produce an intense moderated
positron beam. One should also note that there are fewer research
reactors than available linac's in the world thus limiting this
approach to a small number of institutions.

The initial reactor based beam being constructed at Brookhaven is a simple magnetic transport design similar to several
existing beams. The positrons will be transported at energies of
up to 5 keV using a field of 50-100 Gauss with two $\vec{E}x\vec{B}$ filters (8)
to remove any straight line path in the beam, thereby shielding
from gamma rays and unmoderated positrons (see Fig. 1).

The Brookhaven system will use an evaporated single crystal
Cu source. Owing to the short half-life of Cu^{64} one needs to
transfer the irradiated source every few days thus requiring a re-
motely controlled source preparation chamber. The source chamber
is housed in a concrete shielding house which provides adequate
shielding for up to 10kCi of Cu^{64}. After irradiation of the Cu
sphere it is removed from the reactor and transferred through an
airlock system into an evaporation crucible (Figure 2).

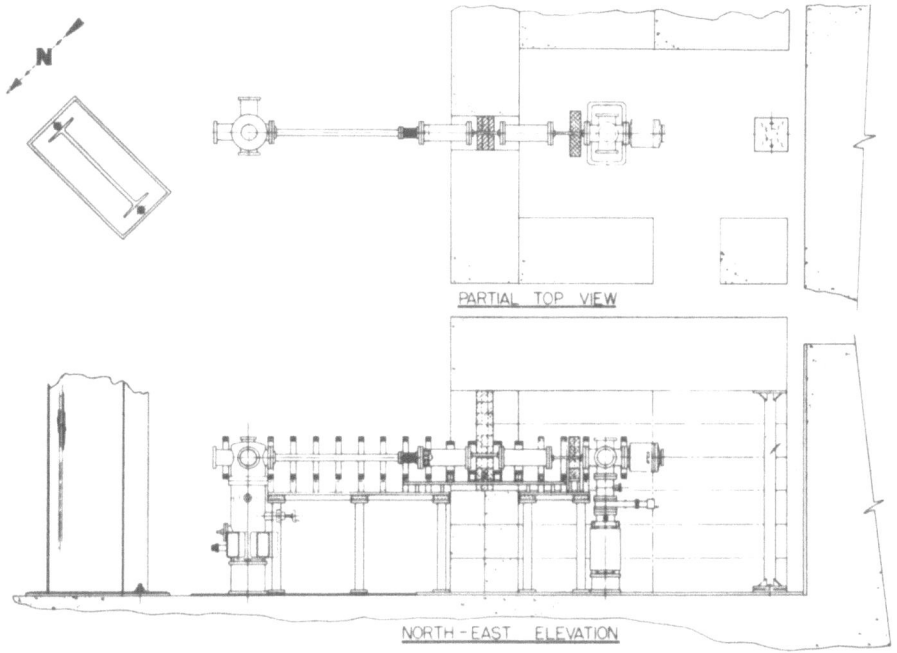

Fig. 1.   The reactor based beam facility being built at Brookhaven.

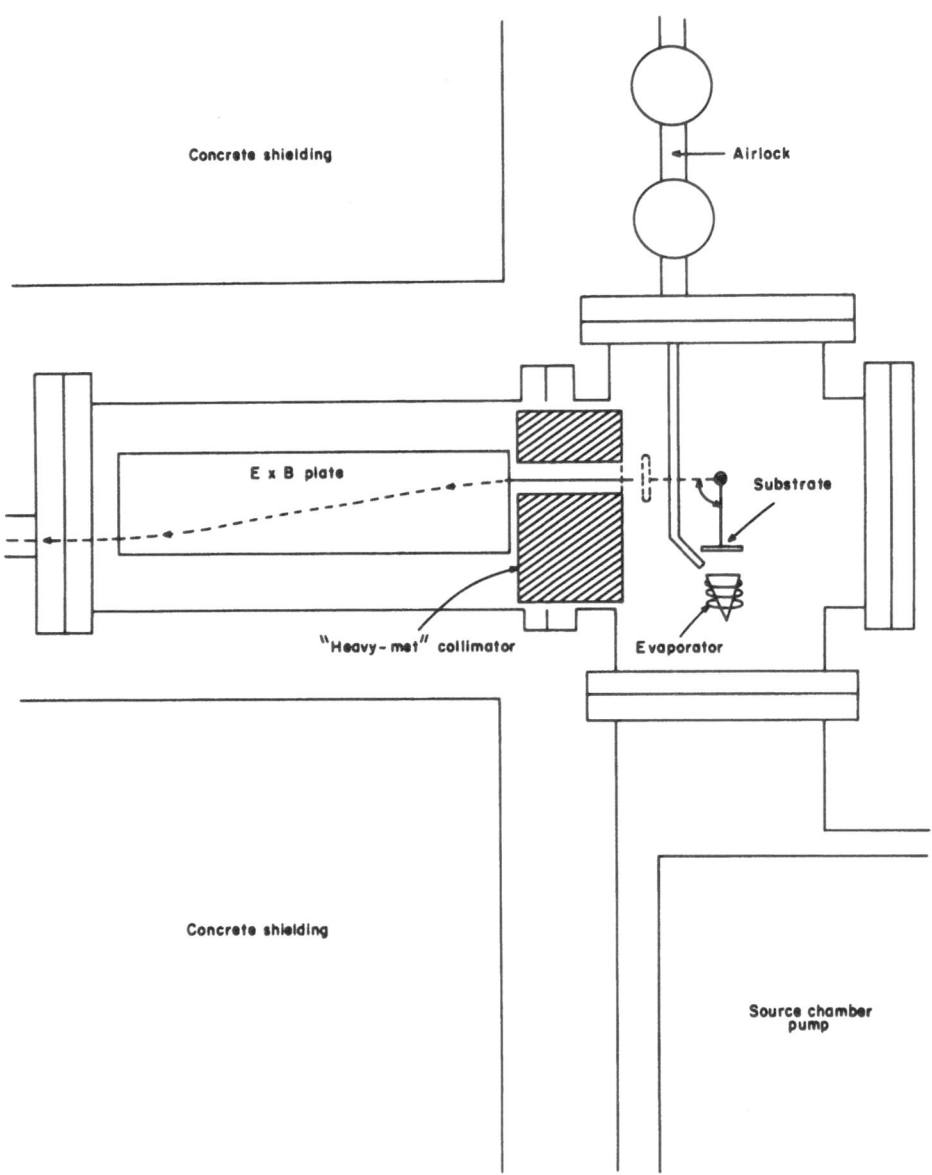

Fig. 2. The source preparation chamber.

A fortuitous property of Cu(111) is that in single crystal form it has been shown to be a very effective positron moderator; therefore, the positron source will be used as its own moderator as well. Those positrons emitted from the Cu^{64} atoms in the bulk of the source/moderator which stop within a few diffusion lengths of the surface ($\approx$1100 Angstroms for Cu) will escape the solid with some energy ranging up to the positron work function (0.6 eV). This self-moderating source can be made by evaporating the irradiated Cu directly onto a W substrate in our source chamber.

It has been found in various measurements that one can grow Cu(111) epitaxially on W(110). Studies have been performed at Brookhaven even at high evaporation rates with good LEED patterns being observed. This indicates that the hybrid W/Cu(111) system should be very effective as a defect free moderator. After the source has decayed one can evaporate another Cu source on top or simply evaporate the Cu off of the W substrate by heating it to high temperatures.

PROPOSED EXPERIMENTS

As we have mentioned, a variety of experiments have been proposed which would benefit from the fluxes available on one of these new machines. One such experiment currently under active consideration at Brookhaven is a study of the reflection of positronium (Ps) atoms from well-characterized solid surfaces.[18] This would provide us not only with a better understanding of Ps-surface interaction but also clarify the possibility of producing beams of positronium atoms for use in the study of surfaces and in particular the possibility of observing Ps diffraction. This subject has been discussed in detail in K. Canter's contribution to this workshop. In the beam proposed at the HFBR one would expect to be able to measure reflection coefficients down to $R=10^{-4}$, using a simple technique. (If Ps generates secondary electrons when impacting a surface this would increase the efficiency of detecting Ps by more than an order of magnitude over a simple annihilation technique, thereby extending our expected sensitivity even further.)

A second experiment planned is to combine the technique of angular correlation of annihilation radiation[19] with the intense slow positron beam.[18] This would provide the first momentum density information about surface electrons sampled positrons bound in the surface states found on most metals. Such an experiment should provide unique information on the momentum density of electrons at surfaces. An additional result would be to probe many-body correlations on the surface. This is undoubtably different from that in the bulk due to the large electric field which is produced by the surface dipole. The first experiments will focus on the simple metals which only have s and p electrons, in

order to more fully estimate and understand these many-body effects. Finally, one could obtain a measurement of the velocity
distribution of emitted Ps by observing the angular correlation
signal of para-Ps atoms which decay very near the surface. This
measurement should provide information regarding the density of
electron states at the surface, microscopic roughness, overlayers
and adsorbates, and many related phenomena.(20)

Other experiments will require the implementation of brightness enhancement of the positron beam such as proposed by
Mills(2). By utilizing this technique one could make a beam with
high spatial and energy resolution using a series of reflection or
transmission moderators. (See A. Mill's contribution in this workshop.) Being somewhat optimistic on these developments one should
be able to reduce the beam diameter by a factor of 50 with about
one-third to one-tenth of the positrons remaining after each stage
of remoderation. A rough estimate with our beam (4×10^8 e$^+$/sec)
yields a reduction to ≈ 1000Å diameter in as few as three stages
with approximately 2% or 10^6-10^7 e$^+$/sec remaining. Another way of
using brightness enhancement would be to reduce the beam diameter
obtained from a relatively large area radioactive source. In the
case of our HFBR beam, a source of 10 cm diameter could yield an
increase of the beam of positrons to 4×10^{10} e$^+$/sec or a current of
6nA of slow positrons.

With positron currents in the nA range one can consider many
new types of experiments. For example, one could produce a low
voltage, high spatial resolution (≈ 1000Å) and easily sweepable
positron beam. This would allow one to build a scanning positron
microscope/microprobe.(1) Such a device would be a unique probe
for studying surface and near surface defects, i.e., dislocations,
vacancies and their clusters, grain boundaries and small particles. One could not only measure the reflected or transmitted
incident beam as in conventional electron microscopes, but in addition, such unique positron characteristics as the energy and
angle of the re-emitted Ps and positrons, and the timing, energy
and angle of the annihilation radiations could all be used as
signals.

One could for example use this high-resolution feature
coupled with the new measurement of re-emitted positron energy
loss spectroscopy(21) to detect whether adsorbed molecules have
specific adsorption sites such as at steps or defects at the surface. This information would be very useful in the field of
catabysis.

Another example would be to couple existing techniques such
as positron lifetime or angular correlation measurements with the
microbeam. The small spot size coupled with a variable beam energy might permit measurement of such characteristics as defect con-

centration as a function of spatial position on a sample. At Brookhaven a higher energy (0-100 keV), poorer spatial resolution (5 mm) beam has been constructed and has shown that one can detect defects in semiconductors caused by ion implantation. Triftshauser and Kogel (22) have seen defects in irradiated metals and are presently coupling their slow positron beam to an ion implantation machine to perform sequential experiments studying the damage created during implantation.

Another potential use for positrons, either in normal or brightness enhanced beams is to remove individual electrons from a sample. Because neither thermalized positrons nor annihilation gamma rays strongly interact with the atoms in a solid, one can view them as a means for abrupt removal of a single electron from its surroundings with a minimum of other disturbance (i.e. no initial or final state complications). One can imagine a number of studies of solids or perhaps of atomic and molecular species in which this would be quite useful. A possibility might be "positron simulated desorption" of adsorbed gas ions. Here positrons should be quite valuable in determining the detailed mechanism by which electron or photon stimulated desorption proceeds(23). A potential benefit in the case of removing electrons in this way is that the electron momentum might well be measureable using the annihilation radiation, thereby even allowing one to be specific as to which electrons have been removed.(24)

An easy measurement one could make using a high flux beam would be to study the time dependence of certain phenomena on a sub-millisecond scale. By measuring the current to a sample under positron irradiation, the number of positrons leaving by re-emission can be determined. By measuring the number of annihilations near the sample in the traditional way one can determine the number of positrons leaving by re-emission or Ps formation. In either case, the use of a high flux beam permits relatively accurate measurements of fast, one-time phenomena. Accuracies of a few percent for a time interval of a few tenths of a millisecond should be possible with rates in the 10^8-10^9 s^{-1} range. One suggestion might be to study the motion of defects near the crack tip in the case of crack propagation. A small diameter, sweepable beam would be an obvious advantage in this case. One could also consider studying surface chemical reactions on this time scale.(25)

Another possibility is to detect positron channeling and to use it as a means of impurity lattice location. For example owing to the positive charge the positron will tend to stay in the channels between the ion cores. If an impurity is located in the channel the positron will be obstructed, thus producing a characteristic blocking pattern or induced impurity x-ray. Positron channeling, in comparison with microbeam proton channeling, gives increased yield and reduced sample damage.

It should be stressed that the complexities imposed by the accelerator or reactor on the experimentor are considerable. When these constraints are coupled with the already complex techniques of the laboratory positron beam, the project becomes extremely formidable. We would suggest that no one embark on the construction of a facility based intense beam without first mastering the techniques of building a laboratory based beam system. This should not disuade anyone, for one should realize that there is a wealth of new condensed matter experiments that can be performed on a laboratory beam. For example, one study has been made on insulating systems and the results are quite surprising and interesting(26). This measurement has shed new light on the formation mechanism of Ps and Ps trapping at defects, and given a value for the Ps diffusion coefficient in a molecular solid. No work has yet been done on positron diffusion in liquid metals or during melting. Preliminary results on the positron diffusion length in semiconductors are very surprising and at the time of this writing not well understood(27).

Very little work has been performed on submonolayer or monolayer overlayers on metal substrates.(28) Quantitative comparisons between the changes in electron and positron work functions in these and other systems are needed to clarify the degree to which these are correlated. Only a few positron work functions have been measured even on clean metal systems (1,29).

Only two preliminary studies have been performed on low energy positron diffraction (30,31) and much work is needed in this area. The spatial dimensions of the positron beam are important to diffraction studies and brightness enhancement would be useful in developing this potentially useful technique.

New work is only now being attempted in measuring the decay rates of positrons associated with the surface of a clean metal. Changes in this lifetime spectra in Al have been detected by changing the surface either by damage or adsorption of an impurity.(14) These positron surface lifetime studies will provide much needed information in unraveling the large amounts of data acquired in bulk positron annihilation studies of voids (internal surfaces) in metals. These voids are generated by neutron irradiation or electron irradiation so that impurities on their surfaces can only be controlled in a very indirect manner(32).

It would be straightforward using a beam to measure the vacancy formation enthalpies of the high melting point metals. Using both Ps and positron re-emission one can make these measurements(1) very rapidly compared to bulk methods. Moreover, bulk techniques themselves can also be performed on a high energy beam as we have mentioned (33).

Very little work has been attempted on measuring the back-scattering fraction of low energy positrons (0-100 keV) on various materials. In fact only a minimal amount of work has been performed on the details of positron energy loss in comparison with electron energy loss in adsorbed overlayers.

New studies utilizing the spin polarization of positrons have been made to measure surface magnetism at various temperatures on a pure metal.(15) Hopefully this work will be continued into thin overlayer systems. These spin polarization studies could also be used for studies in atomic physics such as positron-atom, Ps-atom and Ps-electron scattering.

As one might imagine this list is far from exhaustive. However it is to be hoped that it might leave the reader with the realization that there is a large number of experiments to be performed both on laboratory beams and on intense beams. We also hope that other researchers will continue to consider new uses of positron beams in the future in many areas of pure and applied physics, chemistry and materials science.

ACKNOWLEDGEMENTS

The authors wish to thank H. Lutz for making valuable comments on an earlier version of this paper as well as A. P. Mills and P. J. Schultz for useful discussions. Work performed at Brookhaven National Laboratory is supported by the Division of Materials Sciences, U.S. Department of Energy, under contract DE-AC02-76CH00016.

REFERENCES

1. A. P. Mills Jr., Comments on Solid State Phys. 10, 173
    (1983); Positron Solid State Physics, Proceedings of the 83rd
    Session of the International Summer School "Enrico Fermi" at
    Vanenna, Italy, 1981, W. Brandt and A. Dupasquier, eds.
    (Plenum, N.Y., 1983) – See contributions of K. G. Lynn and
    A. P. Mills, Jr.

2. A. P. Mills Jr., Appl. Phys. 23, 189 (1980).

3. A. P. Mills Jr., Science 218, 335 (1982).

4. L. D. Hulett, J. M. Dale and S. Pendyala, Surf. and Interface
    Analysis 2, 204 (1980).

5. A. Vehanen, K. G. Lynn, P. J. Schultz and M. Eldrup, to be
    published Appl. Physics.

6. H. Jorch, T. E. Jackman, P. J. Schultz and K. G. Lynn,
    unpublished data.

7. K. G. Lynn and H. Lutz, Rev. Sci. Instrum. 51, 7 (1980).

8. A. P. Mills Jr., Appl. Phys. Letts. 35, 427 (1979).

9. T. S. Stein, W. E. Kauppila and L. O. Roellig, Rev. Sci.
    Instru. 45, 951 (1974).

10. J. S. Greenberg and E. D. Theriot, Jr., Vol. IVA, Methods in
    Experimental Physics, Eds. V. W. Hughes and H. L. Schultz.

11. R. H. Howell, R. A. Alverez, K. A. Woodle, S. Dhawan, P. O.
    Egan, V. H. Hughes, M. W. Ritter, Seventh Conf. on the
    Application of Accelerators in Research and Industry, North
    Texas State University, Denton, Tx. (1982).

12. G. Gräff, R. Ley, A. Osipowicz and G. Werth, this workshop.

13. M. Begemann, G. Gräff, H. Herminghaus, H. Kalinowsky and R.
    Ley, to be published in Nucl. Instr. and Methods.

14. K. G. Lynn, W. E. Frieze and P. Schultz to be published.

15. D. W. Gidley, A. R. Koymen and T. W. Capehart, Phys. Rev.
    Lett. 49, 1779 (1982).

16. A. P. Mills, Jr., L. Pfeiffer and P. M. Platzman, to be
    published.

17. Peter J. Schultz, K. G. Lynn, and W. E. Frieze, to be published.

18. This experiment has been proposed with a consortium of researchers which include L. Roellig, S. Berko, K. Canter, A. P. Mills, Jr., W. E. Frieze and K. G. Lynn.

19. See review by S. Berko in Ref. 1.

20. A similar experiment is also being attempted by the Livermore Group on their LINAC based beam – private communication, R. Howell.

21. D. A. Fischer, K. G. Lynn and W. E. Frieze, Phys. Rev. Letts. 50, 1149 (1983).

22. W. Triftshauser and G. Kogel, Phys. Rev. Letts. 48, 1741 (1982).

23. M. L. Knotek, V. O. Jones and Victor Rehn, Phys. Rev. Letts. 43, 300 (1979).

24. K. G. Lynn, J. R. MacDonald, R. A. Boie, L. C. Feldman, J. D. Gabbe, M. F. Robbins, E. Bonderup, and J. Golouchenko, Phys. Rev. Letts. 38, 241 (1977).

25. R. J. Wilson, private commmunication.

26. M. Eldrup, A. Vehanen, P. J. Schultz, and K. G. Lynn, to be published.

27. H. Jorch, K. G. Lynn and I. K. MacKenzie, Phys. Rev. Letts. 47, 362 (1981).

28. P. J. Schultz, K. G. Lynn, W. E. Frieze, and A. Vehanen, Phys. Rev. 1327, 6626 (1983); K. G. Lynn, Phys. Rev. Letts. 44, 1330 (1980); D. Gidlay, private communication.

29. C. A. Murray, A. P. Mills, Jr. and J. E. Rowe, Surf. Sci. 100, 647 (1980).

30. I. J. Rosenberg, A. H. Weiss, and K. F. Canter, Phys. Rev. Lett. 44, 1139 (1980).

31. A. P. Mills, Jr. and P. M. Platzmann, Solid State Comm., 34, 541 (1980).

32. Positron Annihilation, Proceedings of the 6th Int. Conf. at
     Arlington, Texas, P. Coleman, S. Sharma, L. Diana, eds.
     (Plenum, N.Y. 1982).

33. P. J. Schultz and K. G. Lynn, to be published.

INTENSE SLOW POSITRON BEAMS - AN EVALUATION OF THE METHODS

M. Charlton

Department of Physics & Astronomy, University College
London
Gower Street, London WC1E 6BT

The methods of producing intense beams of slow positrons are
discussed and compared. It is shown that, at least for angle
resolved atomic scattering experiments some form of brightness
enhancement is needed to produce the necessary well collimated
beams. The pulsed features of the positron beams produced by
electron LINACS are seen, in many applications, to be disadvan-
tageous.

(I)   INTRODUCTION

The development and application of intense low energy positron
beams will be of special importance in positron physics in the near
future. Over the last ten years steady progress has been made in
improving the efficiency, ε, with which slow positron beams have
been produced starting from the β^+ decay spectra of radioactive
isotopes. Landmarks in the development of non-UHV converters include
the discovery of the MgO converter ($\varepsilon \approx 3 \times 10^{-5}$) by Canter et al.
(1) and the annealed W-vane converter ($\varepsilon \gtrsim 10^{-4}$) of Dale et al.(2).
The mechanism by which slow positrons are ejected from the poorly
characterised MgO and annealed W surfaces is still not fully understood. A more detailed insight into the phenomenon of slow positron
emission was provided by the work of Mills et al (3) and Lynn and
Lutz (4) who, using slow positron beams, studied re-emission from
clean single crystal metal targets under UHV conditions. These
advances resulted in the development of the Cu(III)+S converter (5)
with an efficiency of $\varepsilon = 1.5 \times 10^{-3}$ yielding beams of approximately
4×10^6 slow positrons per second from a 500 mCi Co^{58} source.

Recently a number of methods have been described (6,7) which allow the production of slow positron beams with intensities in the nano-amp region. In one of these cases Mills (6) has described the technique of brightness enhancement whereby the positron beam spot size is progressively reduced by accelerating and focussing a slow positron beam on to successive thin single crystal converters. A single stage of brightness enhancement will allow a reduction in the product of the beam diameter and angular divergence by a factor of $\sim 600 - 1000$ if it is assumed that there is near normal ($\sim 5^{\circ}$) re-emission from a flat target sample bombarded with positrons with energies of a few keV. As discussed by Canter and Mills (8) the application of slow positron beams to angle resolved studies, whether in surface or atomic physics, will require well collimated high flux beams. It will be shown below that all the possible methods of producing intense slow positron beams will, without resorting to aperturing, require at least one stage of brightness enhancement before certain experiments are feasible.

These methods are discussed below with a view to identifying
(a) The typical beam strengths attainable and beam characteristics

(b) The strengths and weaknesses of the various approaches and

(c) The likely costs involved.

Applications of slow positron beams are discussed in section (3). Summaries of beam characteristics, intensities and costs can be found in Tables 1 and 2.

(II) METHODS OF PRODUCTION AND BEAM CHARACTERISTICS

(A) Sources

(1) Conventional Sources

Before discussing the methods which can be applied to produce high yields of slow positrons it will be instructive, for comparison, to consider the conventional source-converter configuration to find the maximum attainable strengths and the beam characteristics. At present the most efficient source-converter configuration is the low self-absorption Co^{58} source in conjunction with the Cu(111)+S converter used in the backscattering mode. A maximum Co^{58} source strength currently commercially available of 500 mCi allows a maximum beam strength of approximately 4.3×10^{6} e^{+} s^{-1}. With the half life of Co^{58} of 71 days this source will have to be replaced 2-3 times annually at a cost of $\sim$ £10,000. Using parameters for the initial average forward kinetic energy, E_1, beam diameter, d_1 and angular divergence, θ_1, limiting beam diameters, d_2 and/or angular divergence, θ_2, can be found at an energy, E_2 using

$$(E_2/E_1)^{\frac{1}{2}} \, m\alpha = 1 \qquad\qquad\qquad (1)$$

where $m = d_2/d_1$ is the linear magnification and $\alpha = \theta_2/\theta_1$ is the angular magnification. Practically it is usually necessary to compute m for a given electrostatic lens configuration before α can be found. In a typical example assuming an initial beam of ($E_1 = 0.5$ eV, $\theta_1 = 30°$, $d_1 = 6$ mm) a 2 mm diameter beam at 20 eV will have an angular divergence of $\backsim 14°$. Such an angular divergence is unacceptably large for many applications and as discussed by Canter and Mills (8) the desired collimation can only be achieved by aperturing the beam. In the above example this could lead to a loss of 90% of the beam intensity.

Under non-UHV conditions the situation is worse. Inserting typical values for the W-vane converter in to equation (1) to produce a beam of 5° angular divergence and a diameter of 2 mm at 20 eV results in a loss of intensity of $\backsim 97\%$ due to aperturing. This situation could probably be improved slightly using a heat treated tungsten converter in the backscattering mode. Experience with the W-vane converter at UCL has shown that yields can vary by factors of $\backsim 4$ using tungsten taken from different parts of a rolled W sheet. Annealing in a methane atmosphere has produced an increase in yield of $\backsim 20 - 30\%$.

Further increases in beam intensity with conventional sources may be possible using a Cu(111) coated W converter. Canter and Mills (8) have quoted the results of preliminary measurements on such a converter at Brookhaven which may have $\epsilon = 10^{-2}$ with a FWHM of the slow positron beam of < 0.3 eV. Maximum beam strength here with a 500 mCi Co^{58} source will be $\backsim 2.9 \times 10^7$ s^{-1}.

An additional gain of a factor of $\backsim 5$, after allowing for a 20% source self-absorption (9), could be achieved by using a Na^{22} positron source. Unfortunately commercially available Na^{22} sources do not allow use with converters in the backscattering geometry. In the future however it may be possible to produce thin Cu(111)+W or single crystal W converters with $\epsilon \backsim 10^{-2} - 10^3$ for use in the transmission mode with Na^{22}. Such an arrangement would have advantages due to the longer half life (2.6 yrs) of Na^{22} as compared to Co^{58} but the cost, given by Lorch (10) as $\backsim 30x$ that for Co^{58} per mCi, could be $\backsim$ £50,000 - £100,000 per source. Lorch (10) also gives the maximum activity of Na^{22} as 400 mCi cm^{-2} which for a 500 mCi source implies a radius of 2 mm. This is approximately 4 times the Co^{58} source spot size and may imply sacrifices in beam brightness.

(2)   Low Energy β^+ Enhancer

In 1970 Lohnert and Schneider (11) described a technique to produce a low energy positron beam by shifting the β^+ energy spectrum from Na^{22} and Co^{58} until the maxima of the spectra were close to zero energy. The principle of their deceleration sphere is shown in Figure 1. The sphere is approximately 100 mm in diameter. Lohnert and Schneider (11) obtained a 'beam' of positrons with energies between 0.5keV and 4keV with an efficiency, $\epsilon = 2.8 \times 10^{-5}$. The low efficiency can probably be attributed to losses during focussing of their 'beam' and West (12) has suggested the use of such a device with a converter to produce a positron beam.

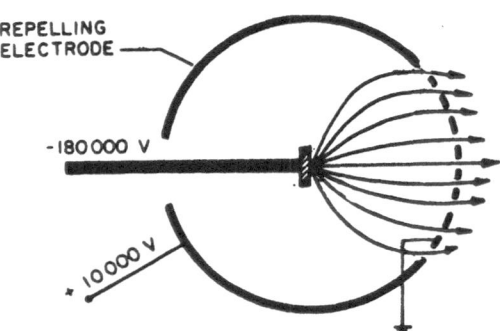

Figure 1:   The deceleration sphere used by Lohnert and Schneider (11)

Assuming that the β^+ losses of Lohnert and Schneider (11) are due solely to focussing, with a retarding potential of 160 keV applied to the source, then $\sim$ 15% of the β^+ can be transported to a mesh in the energy range 0 - 40 keV over an area of $\sim$ 40 cm^2. The highest efficiences will be obtained if geometrical problems are overcome and the positrons produced in this manner are allowed to impinge upon a converter in the backscattering mode. Mills (13) has shown that the slow positron yields, y(E), at a bombarding energy, E, can be written as

$$y(E) = y_0(1 + E^n/E_0^n)^{-1} \qquad (2)$$

where y_0 is the zero energy limit of y(E), E_0 is a converter dependent constant and n = 1.6. Inserting values for the Ni(100) surface given by Mills (13) (E_0 = 8 keV, y_0 = 0.39) we find that the average value of y(E) is $\sim$ 10%. Combining the two efficiences we have

Table 1: Comparison of various source-converter configurations.

Source Converter	Efficiency and Maximum Beams	Costs per annum (source only)	Beam Characteristics (d, θ, E)
Conventional 500 mCi Co^{58} Cu(111)+S	1.5×10^{-3} Approx. 4.3×10^{6} $e^{+} s^{-1}$	$\sim £10,000$	Initial beam $d \sim 6$ mm $\theta \sim 30^{o}$ $E \sim 0.5$ eV
500 mCi Co^{58} Cu(111)+S+W	$\varepsilon \sim 10^{-2}$ (conjectured) $2.9 \times 10^{7} e^{+} s^{-1}$		
Conventional 500 mCi Na^{22} Cu(111)+W+S or W	$\varepsilon \sim 10^{-2}$ (?) $1.5 \times 10^{8} e^{+} s^{-1}$	$\sim £20,000$	As above d may be larger
Enhancer 500 mCi Na^{22} + Ni(100) or other suitable surface	$\varepsilon \sim 2 \times 10^{-2}$ (?) $3 \times 10^{8} e^{+} s^{-1}$	As above	Initial beam $d \sim 70$ mm $\theta \sim 30^{o}$ $E \sim 0.5$ eV
Cu^{64} Reactor produced Typically 7.4 kCi Cu^{64} evaporated as Cu(111)	$\varepsilon \sim 0.5 \times 10^{-3}$ (after 1 stage of brightness enhancement) Approx. $2.8 \times 10^{10} e^{+} s^{-1}$	$\sim £45,000$	Initial source $d \sim 360$ mm $\theta \sim 30^{o}$ $E \sim 0.5$ eV After 1 stage of brightness enhancement d reduced to ~ 3 mm.

$\varepsilon \sim 1.5 \times 10^{-2}$ for the enhancer-converter arrangement. This is
probably an optimistic maximum considering that the converter will

need to have an area equal to or greater than that of the high energy
positrons (~ 40 cm^2). The resulting beam will have a diameter of
approximately 70 mm which, using equation (1), can be focussed down
to a 5 mm diameter without a change in angular divergence at about
200 eV provided a lens system with a linear magnification of 0.07
can be designed. Such a beam may find application in some area of
surface or near surface physics but will be practically useless for
most atomic scattering experiments unless brightness enhancement is
employed.

(3) Cu^{64}

Mills (6) has described a technique whereby high flux positron
beams can be obtained at nuclear sites by the production of large
Cu^{64} sources. This method is technically the most demanding of those
proposed for high flux beams relying for its success upon brightness
enhancement and remote handling of hot radioactive sources into UHV
environments. Cu^{64} sources can be produced from natural copper by
the reaction

$$Cu^{63} + n \rightarrow Cu^{64} \quad .$$

The yield of Cu^{64} is governed by the equation

$$dN(\text{Cu-64})/dt = N(\text{Cu-63})\sigma_1\phi - N(\text{Cu-64})\sigma_2\phi - N(\text{Cu-64})\lambda \qquad (3)$$

where N(Cu-64) and N(Cu-63) are the numbers of Cu^{64} and Cu^{63} atoms
present, σ_1 and σ_2 are the neutron capture cross sections for Cu^{64}
and Cu^{63} respectively, ϕ the thermal neutron flux and λ is the decay
constant of the Cu^{64}. N(Cu-63) can be regarded as a constant due to
the short irradiation times and low capture cross sections. If the
capture of neutrons by Cu^{64} is ignored then equation (3) can be
solved giving

$$\text{specific activity} = \frac{0.6\phi \; \sigma_1(1-e^{-0.693t/T})}{3.7 \times 10^{10} \; W} \; \text{Cig}^{-1} \qquad (4)$$

Inserting values for the activation cross section (4.5 barns) the
atomic weight of the target, W, the Cu^{64} half life, T, (12.7 hrs) with
an irradiation time, t, of 48 hrs (corresponding to 93% of maximum
attainable strength) we find a specific activity of 74 Cig^{-1} per
10^{14} n$_0$ cm^{-2} s^{-1} remembering that the natural abundance of Cu^{63} is
69%. This value is somewhat less then the ~ 6 kCig^{-1} quoted by
Mills (6) at a thermal neutron flux of 10^{15} n$_0$ cm^{-2} s^{-1}.

Assuming the thermal neutron flux given above with a sample of
10 g of natural copper then a source activity of 7.4 kCi will be
produced. The resulting slow positron beam after one stage of

brightness enhancement will be $\sim 2.6 \times 10^{10}$ e^+ s^{-1} after allowance for the Cu^{64} β^+ fraction of 0.19. To produce the 10 Ci slow positron beam strength quoted by Canter and Mills (8) will require thermal neutron fluxes an order of magnitude higher. Beams produced after one stage of brightness enhancement can have a diameter, $d \sim 3$ mm and an angular divergence of $\theta \sim 5^{\circ}$ at 20 eV and would be useful for most crossed beam scattering experiments in atomic physics.

At Harwell in the U.K. the maximum thermal neutron flux is 2×10^{14} n_o cm^2 s^{-1} reducing the useful e^+ beam to about 0.8 nA. It is possible that higher neutron fluxes will be available at other reactor sites in Europe. Irradiation costs at Harwell would be $\sim$ £450 per source and for high flux applications this would need to be replaced every 2/3 days.

(B) Machines

It is now 15 years since the first slow positron beam was produced using the fast positron spectra obtained from an electron LINAC. Groce et al (14) bombarded a tantalum target with a 55 MeV pulsed beam of electrons and using a low efficiency ($\varepsilon \sim 10^{-6}-10^{-7}$) gold converter obtained a beam of a 'few positrons per second'. This method has recently been revived by Howell et al (7) at Lawrence Livermore Labs. U.S.A. and is considered below.

A further method of producing slow positron beams using a machine was developed by Stein et al (15) who used a 10 µA, 4.5 MeV proton beam to activate a boron target via the reaction $B^{11}(p,n)C^{11}$. This technique has been used successfully over the years to measure positron total scattering cross sections (16) with a slow positron beam of narrow energy width (~ 0.1 eV) originating from the boron itself. The C^{11} source strengths produced are limited by the removal of heat from the boron target which in turn restricts the maximum proton beam. This technique will not be treated further.

In the discussion which follows the machines which could be used to produce intense slow positron beams have been divided in to those which produce a pulsed output and those which give a continuous positron beam.

(1) Pulsed Machines

These machines include LINACS, microtrons, synchrotrons etc. which rely upon the use of r.f. accelerating fields and phase focussing to produce a pulsed burst of high energy electrons. Typical pulse widths vary from 5 µs down to 5 ns with repetition rates of the order of 100's - 1000's per second.

The feasibility of using these machines to produce intense slow positron beams has been re-established by Howell et al (7) using a 120 MeV electron LINAC. Employing a 1.2 cm thick tantalum electron-

positron converter coupled with a W-vane fast positron to slow positron
converter Howell et al (7) were able to produce slow e^+ from fast e^-
with a maximum efficiency of $\varepsilon = 1.5 \times 10^{-6}$. * With an average machine
current of 92 μA arriving in 20 ns bursts at 1440 pulses s^{-1} a beam of
$\sim 8.6 \times 10^8\ e^+\ s^{-1}$ was produced. Such primary beams can be measured
directly using a Faraday cup-electrometer system. However with
$\sim 6 \times 10^5$ positrons per machine pulse such rates are far too high to
be measured using any single particle detector. In fact, with pulse
widths in the nano-second range the positron beam produced from the
LINAC can only readily be applied to situations where the signal
levels are less than one e^+ per machine burst.

A similar machine to the Livermore LINAC exists at Harwell.
Again taking an electron pulse width of 20 ns with a mean beam
current of 120 μA at 2000 pulses s^{-1} and using the efficiency of fast
e^- to slow e^+ conversion at the loaded beam energy (87 MeV) we find
$\sim 2 \times 10^9$ positrons per second. At present the LINAC at Harwell is
used for neutron scattering studies and would probably be operated
in a multiplex mode at a hiring cost of around £250 per hour!

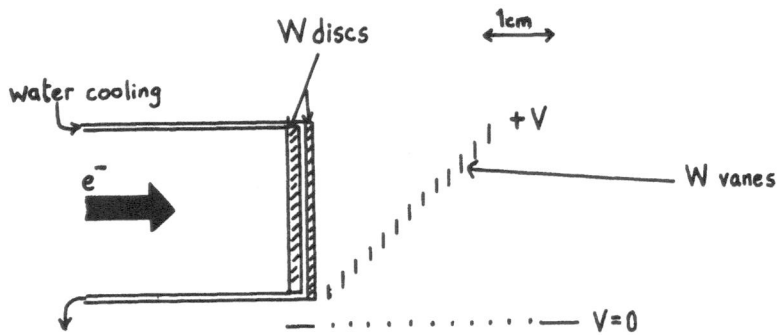

Figure 2: The fast e^--slow e^+ converter in use at Barts.

Due to the high costs of hiring the Harwell machine and the complications involved in using such a machine preliminary experiments
have been carried out using a 15 MeV LINAC at St. Bartholomew's
Hospital Medical School, London, in order to establish the feasibility
of using smaller machines to produce slow positron beams. It is
possible that similar efficiencies to those obtained by Howell et
al (7) can be produced using thinner targets at lower electron beam
energies since the higher the mean energy of the positrons produced
from the e^- - e^+ converter the lower the probability of producing

*   Recently Graaf et al (26)(private communication from Dr. R Ley)
have produced $\varepsilon = 1.1 \times 10^{-4}$ using a 200 MeV electron LINAC at Mainz.

a slow positron. The converter geometry used in the Barts. experiment is illustrated in Figure 2. The e^- beam at Barts. has a mean current of 15 μA with pulse widths > 500 ns repeated at 100 Hz.

Maximum slow positron beam strengths should be around 10^6 per second. At the time of writing this experiment has not been completed.

Table 2: Machines

Method	Efficiency and Maximum Beams	Cost	Beam Characteristics (d, θ, E)
e^--LINAC	$\varepsilon \sim 1.5 \times 10^{-6}$ (7) at e^- energy of 120 MeV. Approximately 8.6×10^8 $e^+ s^{-1}$ $\varepsilon \sim 1.1 \times 10^{-4}$ (26) at e^- energy of 200 MeV Approximately 5.3×10^9 (max. at detector).	Harwell (Hire) £250/hr	PULSED variable beam width Initial beam $d \sim 20$ mm $\theta \sim 30^\circ$ $E \sim 2$ eV.
R.T.M. (Mainz)	$\varepsilon \sim 1.5 \ 10^{-6}$ at 100 MeV (100 μA e^- beam) Approximately 10^9 $e^+ s^{-1}$	?	Continuous Similar initial beam geometry to above.

As mentioned above, however the applicability of pulsed e^+ beams is limited by the lack of a detector able to time resolve events due to single positrons within each burst. This constraint rules out all annihilation studies (at surfaces) and most atomic scattering applications unless detectors can be accurately calibrated to work in a proportional rather than a single counting mode.

Further considerations are the characteristics of the positron beams produced from such machines. The typical beam diameter (given by the area of the W-vanes) is ~ 20 mm which compares unfavourably with beam diameters produced from conventional source-converter geometries. Producing a low energy beam (~ 20 eV) for crossed beam atomic scattering work will require a reduction in beam strength by aperturing to $\sim 2.5 \times 10^{-3}$ of the full amount available. An alternative would be to use the brightness enhancement technique. Up to the present LINAC produced beams have only been used in non-UHV environments. Application of a LINAC to a beam produced in a planar backscattering geometry of the type used in UHV with the Cu(111)+S converter would probably result in a loss of efficiency although it may be possible to use a planar geometry in the transmission mode. Typical beam characteristics are summarised in Table 2.

(2) D.C. Machines

Positron beams produced from these machines would not suffer from the same disadvantages with regard to application as machines working in the pulsed mode. Such machines could include van de Graaf accelerators, superconducting electron LINACS and the three-stage race-track microtron (RTM) described by Herminghaus et al (17). Up to the present time the RTM has only been developed up to the first stage providing a 14 MeV e^- beam which can be used at a current of $\sim$ 15 μA.

This machine has been used by Begemann et al (18) to produce a beam of 2.2×10^5 e^+ s^{-1} using a MgO coated annealed W fast e^+-slow e^+ converter with a fast e^--slow e^+ conversion efficiency of $\varepsilon = 2.3 \times 10^{-9}$. This factor can probably be increased by $\sim$ two orders of magnitude by making improvements in the fast e^+-slow e^+ converter geometry and by removing the MgO coating from the annealed tungsten. Further improvements in slow e^+ yield would probably be obtained by using the second stage of the RTM with electrons of 175 MeV energy with a maximum planned current of 100 μA. It should then be feasible to obtain efficiencies similar to or better than those found by Howell et al (7) ($\sim 1.5 \times 10^{-6}$) giving a continuous slow positron beam of $\sim 10^9$ s^{-1}. It is possible that some extra gains can be made by using the third stage of the RTM which is planned to produce 820 MeV electrons.

Similar considerations to those discussed in section (II)(1) apply with respect to converter geometries and beam characteristics for positron beams produced using these machines and brightness enhancement will have to be developed before angle resolved studies can be contemplated.

(III) APPLICATIONS

A brief discussion of some of the main practical applications of slow positron beams will be given in relation to the above discussion of intense beams.

(1) Solids and Surface Experiments

For certain applications using annihilation radiation detection practical limits are imposed on the intensity of the positron beam. For experiments involving Doppler broadening, measurement of surface state lifetimes and Ps formation and e^+ re-emission studies using gamma ray detectors then the maximum useful beam strengths will be (depending on detectors and geometries) approximately 10^6 - $10^7 s^{-1}$ Such beams can be obtained using conventional source-converter techniques. 2-d angular correlation studies with resolutions similar to those achieved by the groups at UEA (19) and Brandeis University (20) need beam strengths of 10^9 s^{-1} or greater and could be contemplated using a LINAC beam. In most of these cases, however, a pulsed beam of the type derived from a LINAC would be unsuitable due to detector saturation. Use of a weak LINAC beam for determination of e^+-surface state lifetimes could be anticipated if one of the micropulses which make up each machine pulse can be separated from the others (21), (22). A timing spread of $\lesssim$ 100 ps would be required for the positron beam at the target in such an experiment.

The second major class of surface experiment are those which rely upon the detection, after scattering, of the primary slow positron or detection of some other emitted secondary particle. Such studies include low energy positron diffraction, secondary emission of electrons, positronium formation, positronium atomic spectroscopy and even the production of positronium 'beams'. All of these experiments would benefit from having the maximum possible beam intensity however, as above in most cases a pulsed LINAC beam would not be advantageous. (An exception may be multi-photon excitation studies in positronium spectroscopy). Typical detectors used for such experiments would be the single or multi-channel electron multiplier. These detectors exhibit a serious loss of gain around 10^3-10^4 counts s^{-1} (S.C.E.M.) (23) and 10^4-10^5 counts s^{-1*} (C.E.M.A.) (24) due to dead time effects and would be unable to resolve more than one event per machine burst.

(2) Atomic Scattering Experiments

With continuous positron beams none of the experiments in this category, except perhaps those involving positronium formation, will be limited by detector saturation. Experiments involving energy and/or angle resolved positron scattering in crossed beam configurations will require beam strengths $\sim 10^8$-10^9 s^{-1} for adequate signal levels ($\sim$ 1-10 s^{-1}) at angular resolutions comparable to those obtained in conventional electron spectroscopy.

It will be possible to perform many 'total' scattering experiments (e.g. determination of ionisation, excitation and positronium formation cross sections etc.) with beam strengths available from conventional source-converter geometries.

As was the case with the applications discussed in section (III) (1) the use of a LINAC beam to measure 'total' cross sections may result in detector saturation. It is possible, however, that the low signal levels expected in many energy and angle resolved experiments may mean that a LINAC beam can be applied in these cases. In addition high resolution time of flight energy loss studies may be possible.

*This figure refers to a uniformly illuminated channel plate.

(IV) CONCLUSIONS

Projections of maximum beam intensities available from a variety of different methods have shown that the highest yields are expected from the Cu^{64} method of Mills (6) (see Table 1). This method is, however, the most technically demanding of all those discussed and requires the source-converter to be in a UHV environment. The Cu^{64} method also relies entirely upon the development of the brightness enhancement technique. The method of brightness enhancement has been discussed recently by Lynn and Wachs (25) and we have shown here that all methods of intense positron beam production (UHV and non-UHV) will require brightness enhancement if best use is to be made of the available slow positron flux.

It has been shown that in most applications a pulsed beam of slow positrons of the type obtained from a LINAC has the disadvantage of producing detector saturation. Development of an intense, continuous beam of slow positrons seems to offer the most attractive and versatile possibilities for the future.

ACKNOWLEDGEMENTS

M. Charlton wishes to thank the S.E.R.C. for the provision of a Postdoctoral Fellowship. Numerous helpful discussions with Prof. TC Griffith, Dr. GR Heyland and Mr. PJ Curry are acknowledged. The author also thanks Dr. FA Smith, Dr. CD Beling and Mr. JT White for a fruitful collaboration and many informative discussions at Bart's. Thanks are due to Dr. R. Ley for communicating results of the experiments at Mainz prior to publication. Prof. KF Canter Dr. AP Mills Jr. and Dr. KG Lynn are thanked for their hospitality during a visit by the author to the USA in 1982.

REFERENCES

1:   KF Canter, PG Coleman, TC Griffith and GR Heyland. Measurement of total cross sections for low energy positron-helium collisions J. Phys. B. 5:L167 (1972)

2:   JM Dale, LD Hulett and S Pendyala, Low Energy Positrons from Metal Surfaces. Surf. and Int. Anal. 2:199 (1980)

3:   AP Mills Jr., PM Platzman, BL Brown. Slow Positron Emission from Metal Surfaces. Phys. Rev. Lett. 41:1076 (1978)

4:   KG Lynn and H Lutz. Slow positrons in Single-crystal samples of Al and Al-Al$_x$ O$_y$. Phys. Rev. B22:4143 (1980)

5:   AP Mills Jr. Further improvements in the efficiency of low-energy positron moderators Appl. Phys. Lett 37:667 (1980)

6:   AP Mills Jr. Brightness Enhancement of Slow Positron Beams Appl. Phys. 23:189 (1980).

7:   RH Howell, RA Alvarez and M Stanek. Production of slow posi-
     trons with a 100 MeV Electron LINAC Appl. Phys. Lett. 40:751
     (1982).

8:   KF Canter and AP Mills Jr. Slow Positron Beam Design Notes.
     Can. J. Phys. 60:551 (1982)

9:   LD Hulett, JM Dale and S Pendyala. The Effect of Source Window
     Material and Thickness on the Intensity of Moderated Positrons
     Surf. and Int. Anal. 2:204 (1980).

10: EA Lorch. Radiation Sources for Positron Annihilation Applica-
     tions. Proc. 5th Int.Conf. on Positron Annihilation (Japan)
     403 (1979).

11: GH Lohnert and RT Schneider. Generation of a Low-energy Posi-
     tron Beam. Nuc.Tech. 10:315 (1971)

12: RN West. Private communication (1983)

13: AP Mills Jr. Experimentation with low energy positron beams.
     Lecture notes for 83rd session of Int. School of Physics
     'Enrico Fermi', Varena, Italy July 14-21 (1981).

14: DE Groce, DG Costello, JW McGowan and DF Herring. Time-of-
     flight Observation of Low-Energy Positrons. Bull.Amer.Phys.
     Soc. 13:1397 (1968)

15: TS Stein, WE Kauppila and LO Roellig. Production of a mono-
     chromatic low energy positron beam using the $B^{11}(p,n)C^{11}$
     reaction Rev. Sci. Inst. 45:951 (1974)

16: WE Kauppila and TS Stein. Positron-gas cross section measure-
     ments. Can.J. Phys. 60:471 (1982)

17: H Herminghaus, A Feder, KH Kaiser, W Manz and H vd Schmitt.
     The Design of a Cascaded 800 MeV Normal Conducting CW Race
     Track Microtron Nuc Inst. Methods 138:1 (1976)

18: M Begemann, G Graff, H Herminghaus, H Kalinowsky, R. Ley.
     Slow positron beam production by a 14 MeV CW electron accele-
     rator Nuc. Inst. Methods 201:287 (1982)

19: RN West, J Mayers and PA Walters. A high-efficiency two-
     dimensional angular correlation spectrometer for positron
     studies J. Phys. E 14:478 (1981)

20: S Berko, M Haghgooie and JJ Mader. Momentum density measure-
     ments with a new multicounter two-dimensional angular corre-
     lation of annihilation radiation apparatus Phys. Letts.
     63A:335 (1977)

21: GS Mavrogenes, C Jonah, KH Schmidt, S Gordan, GR Tripp and
     LW Coleman. Optimization of isolated electron pulses in the
     picosecond range from a linear accelerater using a streak
     camera-TV diagnostic system. Rev.Sci.Inst. 47:187 (1976)

22: H Kobayashi, T Ueda, T Kobayashi, S Tagawa and Y Tabata
    Performance and Improvements of an Electron Accelerator
    Producing a Picosecond Single Electron Pulse Nuc. Inst.
    Methods 179:223 (1981)

23: Mullard Technical Bulletin 16. Single Channel Electron
    Multipliers (1975)

24: Hamamatsu Technical Manual Res-0795 Characteristics and
    Applications of Microchannel Plates.

25: KG Lynn and A Wachs. Positron Re-emission Brightness
    Enhancement Method. Appl. Phys. A29:93 (1982)

26: G Graff, R Ley, A Osipowicz, G Werth and J Ahrens. Intense
    Source of Slow Positrons from Pulsed Electron Accelerators
    (Private communication from Dr. R Ley, 1983)

THE GENERATION OF MONOENERGETIC POSITRONS

L. D. Hulett, Jr., J. M. Dale, P. D. Miller, Jr.,
C. D. Moak,[a] S. Pendyala,[b] W. Triftshäuser,[c] R. H.
Howell,[d] R. A. Alvarez[d]

[a]Oak Ridge National Laboratory, Oak Ridge, Tennessee
 USA
[b]State University of New York, College at Fredonia
 Fredonia, New York, USA
[c]Hochschule der Bundeswehr, Munich, FRG
[d]Lawrence Livermore National Laboratory, Livermore
 California, USA

ABSTRACT

Many experiments have been performed in the generation and
application of monoenergetic positron beams using annealed
tungsten moderators and fast sources of ^{58}Co, ^{22}Na, ^{11}C and
LINAC bremstrahlung. This paper will compare the degrees of
success from our various approaches. Moderators made from both
single crystal and polycrystal tungsten have been tried. Efforts
to grow thin films of tungsten to be used as transmission
moderators and brightness enhancement devices are in progress.

INTRODUCTION

In 1974 Pendyala[1] reported spectra for positrons emitted
from several different transition metals, among these were
results for tungsten. In 1980 Dale, Hulett, and Pendyala[2] showed
that tungsten could be made into a highly efficient positron
moderator if it is annealed in vacuum at temperatures greater
than 1500°C. Annealing accomplishes two purposes, the cleaning
of the surface by the volatilization of oxides, and the reduction
of defects that decrease the escape probability of the thermalized positrons. Tungsten surfaces prepared by annealing are very
stable in air at room temperature, the oxide coating that forms
is usually less than five monolayers. Thus, the tungsten

195

moderator is very 'practical'; in situ preparation, although
probably desirable, is not necessary; the moderator can be
alternately exposed to air and replaced under vacuum without
serious reduction in efficiency. The annealed tungsten moderator
has been adopted by a large number of other workers doing
experiments with monoenergetic positrons. For example, it was
used in the first low energy positron diffraction experiments, by
Rosenberg, Weiss, and Canter[3] Zitzewitz and co-workers[4] have used
it to obtain high yields of polarized positrons.

Since the initial discovery of the effectiveness of annealed
tungsten as a positron moderator, the authors have made many
attempts to improve its efficiency. Single crystal, polycrystal,
and fine particle tungsten, annealed under varying conditions,
have been prepared and mounted in several different orientations
with respect to the fast source. We have also experimented with
several different fast positron sources, and have made attempts to
study the effects of altering the spectrum of the fast positron
source with absorbing films of different materials. With the
exception of the LINAC work, these results will be presented and
discussed below. LINAC work will be presented by Howell and
Alvarez and their co-workers in another paper.

RESULTS AND DISCUSSION

Annealing Procedure

Annealing of tungsten should be done at temperatures greater
than 1500°C, above the melting point of tungsten oxide. This
effects volatilization of the oxide and cleaning of gross contami-
nants from the surface. For polycrystalline material appreciable
grain growth also occurs, which serves to remove defects that trap
thermalized positrons and therefore increases the distance over
which they can diffuse and escape from the inside of the moderator. The authors have initially reported,[2] that for polycrystalline material, best yields of moderated positrons were obtained
when annealing was done at 2200°C. Since that time we have
annealed at temperatures as high as 2600°C; yields were slightly
better, but only by factors of 50% or less. Vacuum conditions for
the annealing are not critical. Initial studies by the authors
were done using vacuum pressures as high as 10^{-4} torr. Good
vacuum conditions and high annealing temperatures appear to
decrease the amount of energy loss in the moderated positron peak.
The experience of others suggests that moderated positron yields
are better if the tungsten is prepared in situ under high vacuum
conditions. Mills[5] reports that yields can be as high as 3 x
10^{-3} for single tungsten crystal moderators prepared in situ.
On the other hand, Zitzewitz and co-workers[4] report yields as high

as 1×10^{-3} for polycrystal tungsten moderators that have
been exposed to air for short periods of time. The best yields
obtained by the authors were about 5×10^{-4}. All of the
authors' work has been done under conditions in which the moderator was exposed to air between the stages of preparation and use.

Moderator Orientation and Form

Mills[6] has explained in detail why the backscatter
arrangement is preferable for the orientation of the moderator.
His positron guns are arranged such that the moderator surface is
normal to the optic axis along which the slow positrons are
extracted. Their sources, which are ^{58}Co mounted on small
needles, are placed directly in front of the moderator surface.
This allows the exploitation of the special emission characteristics of moderated positrons: the negative work function of the
moderator causes the positrons to be ejected in a direction almost
normal to the moderator surface, therefore the positrons are projected parallel to the optic axis of the gun, allowing them to be
accelerated and condensed into a small spot size. Whenever thick
moderators are used, and the size and shape of the fast positron
source permits, the backscatter configuration is obviously the
preferred arrangement. For ^{22}Na sources, however, such as
that available to the authors, the bulkiness of the capsule makes
this arrangement difficult. We were forced to perform most of our
experiments with the moderator surfaces oriented parallel to the
optical axis of the slow positron gun. For polycrystal moderators
the vane configuration was used in most experiments. That is, the
polycrystal material was cut into strips, 1-3 mm in width, and
mounted as a "venetian blind" arrangement in front of the fast
source. Spacings between the vanes were usually about equal to
their widths. Since the moderated positrons are expected to be
emitted mostly normal to the vane surfaces, they must be turned
90° in direction in order to extract them from the gun. This is
obviously an undesirable situation, leading to poor brightness of
the gun, but it was necessary because of the shape and size of the
fast positron source.

Single crystal surfaces of tungsten, with (110) orientations
have been prepared and arranged in the form of a rectangular
cavity, 6 mm square, 14 mm in length and annealed at 2600°C. The
^{22}Na source was placed at one end of the cavity and the
moderated positrons were extracted from the other end by an
electrostatic field. Experimental arrangements have been shown
diagramatically in a previous report.[7] Slow positron yields were
about the same as for the polycrystalline vane moderators.

Mills[6] has suggested that fine particle moderators, having
larger surface areas, might yield larger quantities of slow

positrons. Two devices of this type, consisting of fine particles
deposited on the inside of cavity configurations, have been
studied. One was prepared by hydrogen reduction of tungsten hexa-
fluoride, particle sizes ranged from 1/2-50 micrometers. Another
was prepared by hydrogen reduction of tungsten oxide smoke, the
average size of its particles was about two micrometers. Neither
of the two fine particle moderators showed an appreciably greater
yield than the polycrystal vane and single crystal cavity configurations. The reason for this is not clear. It is true that
the particle sizes were much greater than the thermalized positron
diffusion lenghts, but one would expect the increased moderator
area to have been beneficial.

The transmission moderator would appear to be the ideal
configuration. It should have two major advantages; a maximum
solid angle of 2π steradians for intercepting the fast positrons
from the source can be obtained; secondly, as the slow positrons
emerge from the front side of the moderator they should be projected parallel to the optical axis of the gun, providing maximum
brightness.

Experiments with three thicknesses of tungsten transmission
moderators have been performed: 25 µm, 12 µm, 2 µm. The 2 µm
moderator was prepared by repeatedly anodizing and stripping the
12 µm material. All three films were annealed at about 1800°C.
In all three cases yields were rather poor, the best being about
20% of that for the vane and cavity configurations. Our
interpretation of these results is that the films were too thick
to serve as effective moderators, and that they utilized positrons
having energies corresponding to relatively small segments of the
fast positron spectrum of the source. Those fast positrons that
are the most effectively moderated probably have the lower
energies. The thick transmission moderators that were used
stopped the lower energy positrons before they reached the front
surfaces, and they came from a more narrow range of the fast
spectrum. We have attempted to measure the lower energy range of
the fast positron spectrum from a ^{58}Co source. The upper
limit of our spectrometer (electrostatic) was 2 keV. No counts
above background could be detected. Apparently the nature of the
surface of the fast source prevents positrons of energy lower than
2 keV from escaping.

It is still our belief that the transmission moderator is the
better configuration, however. Pendyala[8] has calculated that if
the transmission moderator is the same order of thickness as the
diffusion length of the thermalized positrons the yield should be
equivalent to that of a back reflection configuration. We have
therefore begun efforts to grow tungsten films with thicknesses of
the order of the diffusion lengths of the thermalized positrons,

50-200 nm. In our initial approaches we are using the chemical vapor deposition technique. We are exposing various substrates to tungsten hexacarbonyl, which decomposes and deposits metallic tungsten at temperatures above 300°C. Tough, ductile films have been obtained for substrates such as tungsten oxide, mica and MgO. We have been unable to form coherent films on NaCl. We have not yet determined whether the coherent films are epitaxial. We have been only partially successful in floating the deposited films free of their substrates; we have yet to test whether or not they can be annealed into free standing moderators.

Experiments with ^{11}C Fast Sources

Stein and Kauppila[9] have reported the generation of slow positrons by the bombardment of ^{11}B with 4.75 MeV protons from a dynamitron generator. The ^{11}C isotope, which is a fast positron emitter, is formed by the following reaction.

$$^{11}B + H^+ \rightarrow {}^{11}C + n$$

$$^{11}C \rightarrow {}^{11}B + e^+$$

The ^{11}B target becomes self-moderating, presumably because of the build-up of carbon during proton bombardment. Miller and co-workers[10] have studied this reaction for proton energies as high as 10 MeV, using the ORNL tandem van de Graaf. Because of the greater penetrating power of the higher energy particles, the production of ^{11}C was more effective; that is, the target was activated both at the high energies and also at the lower energies of the beam as it became degraded in the penetration of the target. Experiments were started using the ORNL accelerator to bombard boron carbide targets fabricated from ^{11}B and ^{13}C, both of which form positron emitting isotopes when activated with protons. Fifty millicurie quantities of activity were produced, which corresponds to the activity of the ^{22}Na source used in other experiments. An annealed tungsten moderator was used. Yields of slow positrons were very poor, more than a factor of 10 smaller than those obtained with the ^{22}Na source. Several experiments were done to check for problems such as moderator contamination; this does not appear to be the cause of the trouble. We suspect that the poor yields might be due to self absorption of fast positrons by the target. The high energy of the bombarding protons cause them to penetrate to large depths of the target before they undergo activation events. The authors have shown[11] that material placed between the fast positron source and the moderator causes the moderated positron to be significantly decreased.

One of the purposes of this conference is to discuss schemes for devising high intensity sources of monoenergetic positrons to be deployed as user facilities. The ^{11}C source would seem to be attractive and practical; we would like to apprise our colleagues of the trouble we have had with this approach and advise that techniques must be devised to overcome the fast positron self absorption problems.

SUMMARY

Over the past four years our group has performed many experiments in the generation and spectrometry of slow positrons. Many appealing ideas have arisen at workshops such as this, and we have investigated several of them. We have experienced failures as well as successes. The purpose of this paper is to apprise our colleagues of both.

The annealed tungsten moderator has proven to be very useful. A large number of other workers have made use of it. It is easy to prepare and is rugged in use, being tolerant of poor vacuum conditions, and even operable in experiments with gasses.

The development of a fine particle moderator, with a greater surface area to allow larger yields, has not yet been accomplished. Whether or not this idea is sound is not clear.

The transmission moderator, if it is to work, will probably require the fabrication of thin films, less than 200 nm in thickness. Chemical vapor deposition shows promise of being a method by which this can be accomplished.

The ^{11}C fast positron source, produced by proton bombardment, has some difficulties that must be overcome before it can be used for high intensity monoenergetic positron facilities.

ACKNOWLEDGMENT

Research sponsored by the Office of Energy Research, U. S. Department of Energy, under Contract W-7405-eng-26 with the Union Carbide Corporation.

REFERENCES

1. S. Pendyala, D. Bartell, F. E. Girouard, J. Wm. McGowan, Phys. Rev. Lett. 33:1031 (1974).

2. J. M. Dale, L. D. Hulett, and S. Pendyala, Surf. & Interface Anal. 2:199 (1980).

3. I. J. Rosenberg, A. H. Weiss and K. F. Canter, Phys. Rev. Lett. 44:1139 (1980).

4. P. W. Zitzewitz, J. C. Van House, A. Rich, A. Koymen, D. W. Gidley, Physics Department, University of Michigan, Ann Arbor, private communication.

5. A. P. Mills, Bell Laboratories, private communication.

6. A. P. Mills, Proc. of the 83rd Sess. of the Intern. School of Phys. "Enrico Fermi", Varenna, Italy, July 14-17, 1981 (W. Brant and A. Dupasquier, eds.) to be published.

7. L. D. Hulett, J. M. Dale, S. Pendyala, Proc. of the Sixth International Conf. on Positron Annihilation, Fort Worth, TX, April 3-7, 1982 (P.G. Coleman, S. Sharma and L. M. Diana, eds.) North Holland, Amsterdam, The Netherlands.

8. S. Pendyala, Thesis (University of Western Ontario) University Microfilms Inc., Ann Arbor, Michigan, USA.

9. T. S. Stein, W. E. Kauppila, and L. O. Roellig, Rev. Sci. Instrum., 45:951 (1954).

10. J. K. Bair, P. D. Miller, and B. W. Wieland, Int. J. of Appl. Radia. Isot. 32:389 (1981)

11. L. D. Hulett, J. M. Dale, S. Pendyala, Surf. Interface Anal. 2:201 (1980).

APPLICATIONS OF INTENSE POSITRON BEAMS

Richard J. Drachman

Laboratory for Astronomy and Solar Physics
Goddard Space Flight Center
Greenbelt, MD 20771 U.S.A.

INTRODUCTION

This is a wonderful opportunity for me as a theorist: being allowed or even encouraged to propose difficult experiments at a workshop including experimentalists! And to have some realistic hope that the experiments can actually be performed is a new experience as well. The new techniques for producing intense, energy-resolved and well-collimated beams of low-energy positrons seem so good that I expect to see a period of very rapid advance in providing data to confront the theory.

In this report I will discuss positron processes occurring in atomic hydrogen, the negative hydrogen ion and helium. The emphasis will be on measurements of interest in astrophysics as well as some of more general interest. I will not, of course, try to discuss the experiments themselves; I know what is likely to happen to a theorist who does something like that.

AN ASTROPHYSICAL SCENARIO

The recent series of observations, from balloons[1] and from the HEAO-3 satellite[2], have shown that there is an intense, time-variable source of annihilation gamma-radiation somewhere in the direction of the center of our Galaxy. I do not wish to discuss the question of whether the source is actually in the Galactic Center or to speculate about whether a black hole or supernova is the source of the positrons giving rise to the radiation; the atomic physics problem concerns the life-story of the positrons as they slow down and finally annihilate. Astrophysics is a subject

for those with much imagination and creativity, but it does
require some substantial input from the laboratory and from the
quantum theorist.

The most interesting observation concerning the Galactic Center radiation is the fact that the annihilation line is very
narrow; its apparent width is about 3.1 keV, but that value probably reflects just the resolution width of the apparatus. From
this narrow line width we would like to learn something about the
conditions existing in the annihilation region. In addition, we
know that there is some less definite information concerning the
three-photon continuum which measures the fraction of all events
which come from positronium (Ps) annihilation. As the positrons
decrease in energy from their initial high value there is a competition at each collision between direct annihilation (which is
negligible), inelastic scattering and Ps formation. In order for
the annihilation line to be narrow, the temperature of the medium
must be low, and we will assume at first that the medium consists
entirely of neutral atomic hydrogen. At a certain energy below
about 100 eV the Ps-formation cross-section begins to rise
rapidly, and eventually nearly all of the positrons will form Ps
atoms. At the low densities expected all these Ps atoms will
annihilate without colliding. Consequently, the gamma-ray line
width will depend on the distribution in velocity of these Ps
atoms.[3] In turn, this distribution depends on the medium-energy
cross-sections for the competing positron collision processes. Of
course, the positron-hydrogen system provides us with the simplest
imaginable atomic collision problem; you might reasonably suppose
that the theory had been completely worked out long ago. I would
like to describe the present status of the theory, and try to con-
vince you that there is much that still needs to be learned. I
hope that the new possibility of crossed positron-atom beam
experiments will fill the blanks in our knowledge and also support
the parts of the theory that seem to be well founded.

The most reliable calculations in all of positron-atom physics involve the elastic-scattering region in hydrogen below 6.8
eV; the first three partial-wave phase shifts are very accurately
known[4], and the higher phase shifts can be well approximated by
the effective-range expression[5]. It would be very desirable to
check this region experimentally, but I would be quite surprised
if any serious discrepancies appeared. On the other hand, much of
the astrophysically interesting energy region above 6.8 eV is only
qualitatively known from theory. In figure 1 the four relevant
cross sections in this region (elastic, excitation,ionization and
Ps-formation) are plotted in approximations that have been used in
the analysis of the Galactic Center radiation[3,6]. The elastic
cross section was obtained from the analytic continuation method[7]
which has been applied only up to an energy of about 30 eV. The
excitation cross section includes only the n=2 levels of hydrogen[8]

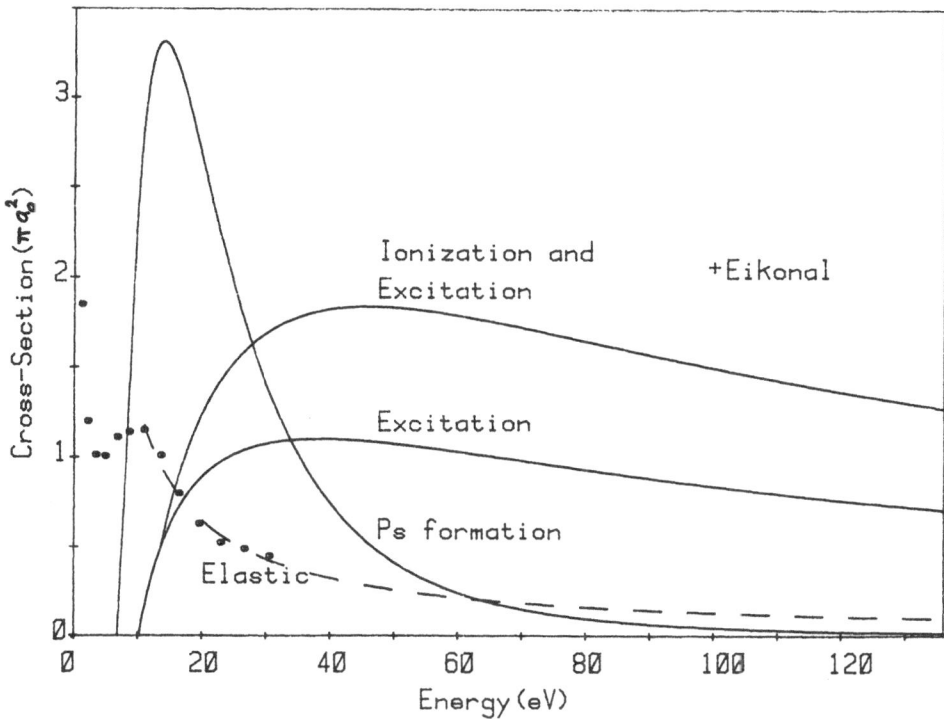

Fig. 1. Cross sections for e^+-H processes.   Circles
below threshold are from Refs. 4 and 5, those above are
from Ref. 7, and the broken line goes like 1/E.   The
cross is from Ref. 10, and the solid curves are smooth
fits to the cross sections as discussed in the text.

and has been fitted analytically for display purposes. the ionization result shown is from the analytic fit to the electron-
hydrogen experiments[9], and its reliability for positrons is
completely unknown. An eikonal approximation at higher energy is
also shown [10]. Finally, the curve for Ps formation, a most
important component in the analysis of the gamma-ray line width,
is taken as a hybrid[11] of the Born approximation for the higher
partial waves and the results of the essentially exact results for
the s- and p-waves[12]. (There are now quite a few new calculations
of this process in a variety of approximations, which I will not
discuss here.) Let me remark that this Ps-formation cross section
does not include excited-state formation, which could be roughly
estimated using a $1/n^3$ law.

To estimate the self-consistency, if not the absolute cor-
rectness, of these cross sections one may apply the forward dis-

persion relations, now known to be correct for positron-hydrogen collisions[13]. Since we wish to emphasize the cross sections in the intermediate-energy region, the following subtracted form is most useful:

$$\text{Re } [f(k) - f(0)] = \frac{k^2}{2\pi} P \int_0^\infty \frac{\sigma(k')}{k'^2 - k^2} dk' \qquad (1)$$

Remarkably, the left and right sides of Eq.1 agree to within 5% except at the lowest value of k=.1; the relative contributions of the cross sections above and below threshold differ considerably as k varies from .1 to .6. There is thus no clear indication of any disagreement with the primitive cross sections that have been used so far in the astrophysical analysis, but an increase in the inelastic cross sections as indicated in the figure might be compensated by a decrease in the Ps-formation value. I have a personal leaning in this direction, not strong enough to be called a theory.

POSITRONIUM FORMATION IN H⁻

The collision of low-energy positrons with negative hydrogen ions in the laboratory would be impossible unless crossed-beam or merging-beam experiments were feasible. There is considerable theoretical interest and some astrophysical interest in this process. On the one one hand, it is of some interest to note that the reaction

$$e^+ + H^- \rightarrow Ps(nl) + H(1s) \qquad (2)$$

is exothermic, and that three distinct Ps atomic levels result from zero-energy collisions: n=1-3. Each of these levels gives a distinct Ps kinetic energy, and astrophysically these three groups of atoms would contribute gamma-ray line components of different widths and intensities. Although the cross sections for these reactions are now under study, I cannot promise accurate results in the near future, and an experimental study would be desirable. I can point out that the cross section for reaction (2) is many orders of magnitude larger than radiative Ps formation on free electrons, and a fairly small fraction of H⁻ ions serves to make this process competitive; nevertheless, only in some fairly special regions (transition regions of planetary nebulae) are high densities of H⁻ ions expected.

The other interesting feature of reaction (2) is its connection with the high Rydberg states of Positronium Hydride (PsH).[14] There is a complete series of hydrogenic levels of the system H⁻ + e⁺ that are stable except for the re-arrangement reaction (2); they represent resonances in the Ps + H scattering channel. For each of these resonances an energy and a width can be computed[14,15], and these parameters can be translated into a complex quantum defect. In turn, the complex quantum defect gives rise to an estimate[16] of the complex phase shift near the threshold for reaction(2). In this way one can hope to check the detailed computations of the Ps formation cross section in H^-. The lowest state of PsH, the 1s state, is bound by more than 1eV, but it has not so far been detected experimentally. In principle, radiative capture of a positron by H^- could produce it, but the rate, even with the high-intensity merged-beam experiments being discussed here, would be unreasonably low. Let me just mention, for completeness, the difficult reaction

$$Ps + H^- \rightarrow PsH + e^- \qquad\qquad (3)$$

which would give rise to the PsH molecule. The energy of the outgoing electron would indicate whether the ground state or one of the resonances had been produced.

ANNIHILATION IN HELIUM

A long time ago I made a simple calculation[17] concerning the final state of the He^+ ion remaining after the annihilation of one of the atomic electrons in neutral helium. To a very good approximation the probability of the ion being left in the nth hydrogenic state is given by the overlap expression:

$$P(n) = \frac{\int d\vec{x}\,\left|a_n(\vec{x})\right|^2}{\int d\vec{r}\int d\vec{x}\,\left|\Psi_k(\vec{x},\vec{x},\vec{r})\right|^2}, \qquad \text{where}$$

$$a_n(\vec{x}) = \int d\vec{r}\,\phi_n^*(\vec{r})\,\Psi_k(\vec{x},\vec{x},\vec{r}). \qquad\qquad (4)$$

Here the scattering wave function of the positron-helium system must be computed and taken at the point where the positron and one of the electrons coincide. From a fairly crude polarized-orbital calculation I found that about 1.2% of all annihilations of zero-energy positrons would leave the daughter ion in the 2p state, when cascading was taken into account, and the Lyman-alpha line would be observable in principle. In addition, about 2% of the annihilations would leave the ion in the metastable 2s state, subject to quenching. I thought it might be interesting to look for the ultraviolet radiation in coincidence perhaps with the annihilation gamma rays.

At that time, there was one very serious obstacle to performing such an experiment: The high-energy incident positrons would flood the apparatus with ultraviolet radiation due to ions pro-

duced during thermalization. The obvious point to be made is that
once one has intense beams of positrons to be injected at an energy below the ionization threshold, it might be possible to observe
the de-excitation radiation described above.

CONCLUSIONS

I have presented a short list of measurements on simple atoms
and ions which I would like to see performed with the proposed
high-intensity low-energy positron beam facilities. Many more
could be proposed, but I hope that these will be seen to be just-
ified theoretically and interesting enough to deserve a fairly
high priority on the experimental queue.

REFERENCES

1. W. N. Johnson and R. C. Haymes, Astrophys. J. 184:103
    (1973); R. C. Haymes et al., Astrophys. J. 201:593
    (1975); F. Albernhe et al., Astron. Astrophys 94:214
    (1981); M. Leventhal et al., Astrophys. J. 240:338
    (1980); M. Leventhal, C. J. MacCallum, and P. D.
    Stang, Astrophys. J. 225:L11 (1978).
2. G. R. Riegler et al., Astrophys J. 248:L13 (1981).
3. C. J. Crannell et al. Astrophys J. 210:582 (1976); R. W.
    Bussard, R. Ramaty, and R. J. Drachman, Astrophys. J.
    228:928 (1979).
4. A. K. Bhatia et al., Phys. Rev.A 3:1328 (1971); S. K.
    Houston and R. J. Drachman, Phys. Rev. A 3:1335
    (1971); A. K. Bhatia, A. Temkin, and H. Eiserike,
    Phys. Rev. A 9:219 (1974); D. Register and R. T. Poe,
    Phys. Lett. 51A:431 (1975).
5. T. F. O'Malley, L. Rosenberg, and L.Spruch, Phys. Rev.
    125:1300 (1962).
6. R. J. Drachman, in "Positron-Electron Pairs in Astro-
    physics," M. L. Burns, A. K. Harding, and R. Ramaty,
    eds., A. I. P., New York (to be published).
7. J. R. Winick and W. P. Reinhardt, Phys. Rev. A 18:925
    (1978).
8. L. A. Morgan, J. Phys. B 15:L25 (1982).
9. W. Lotz,Z. fur Physik 206:205 (1967).
10. F. W. Byron, Jr, C. J. Joachain, and R. M. Potvliege,
    J. Phys. B 15:3915 (1982).
11. R. J. Drachman, K. Omidvar, and J. H. McGuire, Phys.
    Rev. A 14:100 (1976).
12. J. Stein and R. Sternlicht, Phys. Rev. A 6:2165 (1972);
    Y. F. Chan and P. A. Fraser, J. Phys. B 6:2504 (1973);
    Y. F. Chan and R. P. McEachran, J. Phys. B 9:2869
    (1976); J. W. Humberston, Can. J. Phys. 60:591 (1982).

13. E. Gerjuoy and C. M. Lee, J. Phys. B 11:1137 (1978); E. G. Drukarev, J. Phys. B 14:L203 (1981).

14. R. J. Drachman, Phys. Rev. A 19:1900 (1979).

15. R. J. Drachman and S. K. Houston, Phys. Rev. A 12:885 (1975); B. A. P. Page, J. Phys. B 9:1111 (1976); Y. K. Ho, Phys. Rev. A 17:1675 (1978).

16. M. J. Seaton, Rep. Prog. Phys. 46:167 (1983).

17. R. J. Drachman, Phys. Rev. 150:10 (1966).

APPLICATIONS OF INTENSE POSITRON BEAMS IN ATOMIC PHYSICS EXPERIMENTS

Günther Sinapius

Universität Bielefeld
Fakultät für Physik
D-4800 Bielefeld 1, Fed. Rep. of Germany

INTRODUCTION

The aim of this talk is to summarize the experiences with different positron sources and to point out which experiments will be possible in the near future. In this context I shall restrict myself to atomic physics experiments.

INTENSE POSITRON BEAMS

Moderators

Both intense sources of fast β^+ and efficient moderators are necessary in order to obtain intense slow positron beams. Scattering experiments in atomic physics operate with target-gas thickness of the order of 10^{15} particles/cm^2. With moderate efforts in differential pumping a basic pressure of 10^{-7} Torr in the apparatus is achieved. Thus moderators as used in surface studies and the brightness enhancement schemes based thereupon cannot be employed for this kind of experiment. Under these experimental conditions tungsten moderators prepared according to Dale et al. (1) give the highest conversion efficiency $\varepsilon = 10^{-3}$; ε is the ratio of the extractable positron intensity to the β^+ activity of the source. The β^+ from the ^{11}C source are moderated by the ^{11}B with an efficiency of up to 10^{-6} (2).

Not all the extractable positrons, however, are available for a scattering experiment. The beam transport system, electrostatic or with magnetic guiding fields, the requirement of a narrow energy width and the finite aperture of the gas cell cut the beam intensity. Let me give an example from our experiment for the measurement of total cross sections (3). The positrons start from the 10 mm diameter venetian blind tungsten moderator with an energy of about 1.5 eV. Due to the venetian blind geometry, the vanes being parallel to the beam direction, the moderated positrons presumably are emitted with a cos Θ distribution. With reasonable electron optics all positrons emitted within a cone of 120 degrees can be extracted. In the experiment the beam has to pass a gas cell with 3 mm apertures with a maximum divergence of ± 3.7 degrees. For measurements at 100 eV positron energy Liouville's theorem leads to a reduction of the beam intensity by

$$100(\sin 3.7^{o} \times 1.5)^{2}/[1.5 (\sin 60^{o}. \times 5.0)^{2}] = 1/30.$$

As a matter of fact, this is still the best case estimate because both imperfections of the optics and the influence of uncompensated magnetic fields will further reduce the beam intensity.

Comparing different moderators rather the brightness per unit energy (4) than the moderation efficiency is relevant. In our scattering experiment a platinum moderator yielded about the same positron intensity as tungsten, whereas Zitzewitz found tungsten to be about five times as efficient as Pt in a beam that accepted a wide energy width (5). Apparently, in our apparatus the high moderation efficiency of the tungsten moderator is partly cancelled by the larger emission energy and energy width (above 1 eV) of the positrons.[+] The boron moderator emits positrons with a measured energy of less than 1 eV and a width of 0.1 eV (2), but its moderation efficiency is difficult to reproduce (3).

Fast Positron Sources

Radioactive Sources. The currently used radioactive sources are ^{11}C, ^{22}Na, and ^{58}Co. Activities of up to 50 mCi for ^{22}Na (5), 300 mCi for ^{58}Co (6) and some hundred mCi for ^{11}C (3) are employed. Whereas for Na and Co costs are the limiting factor (7), the ^{11}C activity is restricted by the heating up of the boron during activation (3,4). For the same reason the active area of ^{11}C has to be larger than of commercial ^{22}Na sources (about 3 mm). Radioactive sources supply up to 10^{10} fast β^{+}/sec. Higher intensities should be feasible with ^{64}Cu sources (7). The highest positron intensity in a scattering experiment reported so far is 5000/sec obtained with a ^{11}C source and moderation in the boron (8). With a 50 mCi ^{22}Na source and a tungsten moderator, Zitzewitz reports 5×10^{6} pos trons/sec (5).

Electron Accelerators. Another way to obtain a high β^+ intensity is to work on electron accelerators. When an electron with an energy E above a critical energy E_c hits a converter with the atomic number Z, radiation loss dominates and subsequently (e+ e-)-pairs are produced. The values for E_c are given by (10)

(1) $$E_c \approx 600/Z \text{ MeV}.$$

The optimum converter thickness and the maximum number of particles in the shower may be deduced from a qualitative description (10) This crude model yields an optimum converter thickness t (in radiation lengths):

(2) $$t = \ln(E/E_c)/\ln 2.$$

In this depth the shower will contain N particles, with photons, fast electrons and positrons approximately equal in number, with

(3) $$N = E/E_c.$$

This method has been successfully employed by Groce et al. (10) and is now being used at Lawrence Livermore Laboratory (11) and at Mainz (13). With a tantalum converter for the electron beam and a tungsten moderator up to 9×10^{-6} positrons per incident electron and up to about 6×10^7 positrons/sec have been observed at the Mainz accelerator. A cooled converter should allow a higher electron current and a further increase in positron intensity. The pulse structure of the accelerator provides time marks to the accuracy of several nanoseconds (with reduced intensity). Because of the radiation at electron accelerators possible experiments have to be set up further away from the moderator. To provide an efficient transport of the positrons over several meters, magnetic guiding fields are mandatory. There will be no problem for experiments which can do with these fields. Extracting the positrons from the magnetic field will lead to a loss in brightness, unless the beam is accelerated onto a second moderator (brightness enhancement).

PROPOSED EXPERIMENTS

With the increased positron intensity the measurement of angular resolved and partial cross sections becomes possible. Assuming that the highest positron intensity will be available at electron accelerators, I start with the discussion of an experiment that can be done in the presence of the magnetic guiding field.

Positron Impact Ionisation

Positron impact ionisation cross sections for helium have been
deduced from calculated cross sections for the elastic scattering
and the most important excitations (14) and comparison with the
measured total cross sections. Just above threshold they are twice
as high as for electron impact ionisation, presumably due to positronium formation. So far only positronium formation has been
studied experimentally (17,18). The energy dependence of the ionisation cross section near threshold is controversial (15,16).

Our group plans an experiment to measure positron impact ionisation. The apparatus is designed to measure the total number of
ions produced. It is to be done at the electron accelerator at
Mainz. The accelerator operates at 100-150 cycles/sec yielding up
to 10^6 slow positrons per pulse. The 200 Gauss guiding field
confines the beam to about 1 cm diameter. Traversing a He target
the positrons will produce ions. The target consists of a 1 cm
diameter 50 cm long tube, operated as a Penning-trap to store the
ions. It will be emptied after each positron pulse. Behind the
target the He^+ ions and the positrons will be separated and
detected with channel-plate detectors. In the middle of the target
a gas pressure of 6×10^{-4} Torr is supplied. It is differentially
pumped from both ends. The target thickness will be about 5×10^{14}
atoms/cm^2. With an ionisation cross section for helium of 5×10^{-17}
cm^2 (14) about 2% of the positrons will ionize a He atom. Accounting
for losses at the entrance of the target we expect a slow positron
intensity of about 10^4/pulse, producing about 200 He^+ ions.

There are several difficulties. Along with the positrons
there will be many more electrons emerging from the moderator and
they will be equally well transported by the magnetic guiding field.
They have to be removed from the beam by appropriate repelling
potentials at the moderator. Because of partially viscous flow
and helical trajectories in the target the effective target gas
thickness is not well-known. For this reason we assume electron
and positron ionisation cross sections to be equal at sufficiently
high energies, say 400 eV. In order to have similar positron trajectories and thus effective target thicknesses at different
energies the magnetic field should be scaled like $E^{0.5}$. The measurement of the positronium formation cross section will be a
possible extension of this experiment. The gammas from the
Ps decay can be counted with a NaI-detector. The discrepancy

between the experimental results (17, 18) suggests that there may
be difficulties to obtain the Ps-formation cross section from the
observed o-Ps decays alone. Within its life time of ~ 142 nsecs
an o-Ps atom with a kinetic energy E will travel $\sim 6 \cdot \sqrt{E(eV)}$ cm.
Thus it may either escape from the detection region or hit a wall.
The two-γ signals from the p-Ps decay will be augmented by the
latter process. The pulse structure of the electron accelerator is
necessary for background supression. The aim is to measure both
contributions to the ionisation cross section, i.e. Ps-formation
and direct ionisation. In a preliminary step the apparatus will be
tested in Bielefeld with a 3 mCi ^{22}Na source. In a simplified
version of the apparatus only direct ionisation can be measured.
Whenever a positron passes the gas target and hits the detector,
the Penning trap will be opened to look for He$^+$ ions. In Bielefeld
the positron intensity will be rather low ($<10^4$/sec) but the experiment can be run continuously, whereas the time schedule at an
accelerator is much tighter.

Differential Cross Sections

     The first study of the differential scattering of positrons
was performed in a magnetic field using the TOF-method (19). Differential scattering experiments, similar to the electron ones, require
about 10^7 positrons/sec (20). We prepare a crossed beam experiment
for the time when sufficiently intensive positron beams will be
available. The molecular beam will be formed by a source consisting of a long tube or an array of tubes. Its intensity and
angular distribution is determined by the length and diameter of
the tubes and the pressure behind the source. There are not many
experimental data on the target thickness achieved with molecular
beams. From the comprehensive experimental study of molecular
beams in (21) we estimate that a target thickness of about 3×10^{12}
particles/cm^2 for a 0.5 cm diameter molecular beam is feasible. Good
pumping provided, a channel-plate detector for the scattered
positrons can be mounted 4 cm away from the beam intersect. In
this configuration a 2.5 cm diameter detector covers a solid angle
of 0.3 sr. For most gases the positron differential cross section
below 100 eV is of the order of 10^{-17} cm^2/sr. In order to detect
ten scattered positrons/sec (which is about equal to the maximum
dark count rate of the channel-plate detector), the required
positron intensity is 10^6/sec. With a 50 mCi source first measurements seem to be possible.

FUTURE EXPERIMENTS

The most challenging experiment to be done is the scattering of positrons on atomic hydrogen. I shall give some estimates on the feasibility of this kind of experiment. There are two ways to produce atomic hydrogen targets: through the thermal dissociation (22) or DC discharge in a Wood's tube (23).

At temperatures above 2000 K molecular hydrogen thermally dissociates to more than 90%. Thus an atomic hydrogen target can be prepared in a high temperature oven. In order to avoid magnetic field effects the heating current has to be gated. The reported target densities range up to 3×10^{12} atoms/cm^3 in a 2.5 cm long target cell (22) which corresponds to a target thickness of 7×10^{12}/cm^2. The gas discharge dissociation device yields a higher molecular background than the high temperature oven. This is due to recombination of hydrogen on its way from the discharge region to the effusive target. In the device described in (23) a degree of dissociation of about 70% and an atomic hydrogen target thickness of up to 5×10^{13}/cm^2 has been achieved. The total cross section for e$^+$-H scattering is below 5×10^{-16} cm^2 (24). A positron beam passing through an atomic H target would be attenuated by less than 2%, making transmission experiments quite impossible. The measurement of partial cross sections, picking out a single reaction channel, seems more promising. For these experiments the geometry of a gas discharge dissociation source seems to be better suited than a high temperature oven.

Recently positronium formation in atomic hydrogen has attracted particular interest because of its implications in astrophysics (25). Theoretical results for the positronium formation in atomic hydrogen yield cross sections in the order 10^{-16} cm^2 close to threshold (26). In a gas discharge target about 0.5% of the incident positrons will form positronium on atomic hydrogen. If 0.1% of the positronium decays are detected, with a positron intensity of 10^7/sec there will be 50 counts/sec. There will be a background due to the molecular hydrogen in the target.

References

1.   J. M. Dale, L. D. Hulett and S. Pendyala, Low energy positrons from metal surfaces, Surf. Interface Anal. 2:199 (1980).
2.   T. S. Stein, W. E. Kauppila and L. O. Roellig, Near-thermal energy width of moderated high energy positrons, Phys. Lett. 51A:327 (1975).

3. A. Deuring, K. Floeder, D. Fromme, W. Raith, A. Schwab,
    G. Sinapius and P.W. Zitzewitz, Total cross section
    measurements for positron and electron scattering on
    molecular hydrogen between 8 and 400 eV, J. Phys. B 16:
    1633 (1983).

4. A.P. Mills, Jr., Experimentation with low energy positron
    beams, Lecture notes for the 83rd session of the
    International School of Physics "Enrico Fermi" (1981).

5. J. Van House and P.W. Zitzewitz, private communication (1983).

6. K. G. Lynn and H. Lutz, Slow-positron apparatus for surface
    studies, Rev. Sci. Instrum. 51:977 (1980).

7. K. F. Canter and A. P. Mills, Jr., Slow positron beam design
    notes, Can. J. Phys. 60:551 (1982).

8. K. G. Lynn and W. E. Frieze, Intense positron beams and possible
    experiments, this workshop.

9. W. E. Kauppila, T. S. Stein, J. H. Smart, M. S. Dababneh,
    Y. K. Ho, J. P. Downing and V. Pol, Measurements of total
    scattering cross sections for intermediate energy positrons
    and electrons colliding with helium, neon, and argon,
    Phys. Rev. A 24:725 (1981).

10. D. H. Perkins, "Introduction to High Energy Physics",
    Addison-Wesley Publishing Company, Reading (1972).

11. D. E. Groce, D. G. Costello, J. W. McGowan and D. F. Herring,
    Time-of-flight observations of low-energy positrons,
    Bull. Amer. Phys. Soc. 13:1397 (1968).

12. R. H. Howell, R. A. Alvarez, K. A. Woodle, S. Dhawan,
    P. O. Egan, V. W. Hughes and M. W. Ritter, Intense
    positron beams: Linacs, this workshop.

13. G. Gräff, R. Ley, A. Osipowicz, and G. Werth, Intense
    source of slow positrons from pulsed electron accelerators,
    this workshop.

14. S. L. Willis and M. R. C. McDowell, Pseudostate effects on
    positron-helium excitation cross sections at intermediate
    energies, J. Phys. B 15:L31 (1982).

15. H. Klar, Threshold ionisation of atoms by positrons, J. Phys.
    B 14:4165 (1981).

16. A. Temkin, Threshold law for positron-atom impact ionisation,
    J. Phys. B 15:L301 (1982).

17. T. C. Griffith, Positronium formation cross-sections in various
    gases, this workshop.

18. P. G. Coleman, this workshop.

19. P. G. Coleman and J. D. McNutt, Measurement of differential
    cross sections for the elastic scattering of positrons by
    argon atoms, Phys. Rev. Lett. 42:1130 (1979).

20. P. A. Fraser, B. H. Bransden, P. G. Coleman and W. Raith,
    Round-table discussion on future directions in positron-
    gas research, Can. J. Phys. 60:565 (1982).

21. J. A. Giordmaine and T. C. Wang, Molecular beam formation by
     long parallel tubes, J. Appl. Phys. 31:463 (1960).
22. R. A. Phaneuf and F. W. Meyer, Single-electron capture by
     multiply charged ions of carbon, nitrogen, and oxygen in
     atomic and molecular hydrogen, Phys. Rev. A 17:534 (1978).
23. B. A. Huber, A. Bumbel and K. Wiesemann, A high-density
     effusive target of atomic hydrogen, J. Phys. E 16:145
     (1983).
24. J. R. Winick and W. P. Reinhardt, Moment T-matrix approach to
     e^{+}-H scattering, II. Elastic scattering and total cross
     sections at intermediate energies, Phys. Rev. A 18:925
     (1978).
25. R. J. Drachman, Positron astrophysics, in: "Positron
     Annihilation", P. G. Coleman, S. C. Sharma, L. M. Diana,
     eds., North-Holland Publishing Company, Amsterdam (1982).
26. P. Mandal and S. Guha, On the first-order approximations for
     positronium formation in atomic hydrogen, J. Phys. B
     12:1603 (1978).

LOW ENERGY POSITRON AND POSITRONIUM DIFFRACTION

Karl F. Canter

Physics Department
Brandeis University
Waltham, MA

Although this workshop is primarily concerned with positron
scattering (and annihilation) with individual atoms in the gaseous
state, it is clear that in order to understand the scattering of
positrons and positronium Ps from solid surfaces one must often
employ concepts directly related to the problem of scattering in
gases. In the case of a monolayer of atoms (or molecules) physisorbed to a surface, the physisorbed layer can be treated accurately as a two dimensional gas of weakly perturbed atoms.[1] Elastic
scattering, for example, of positrons and Ps from such an adsorbed
layer would then indirectly contain information about the scattering from the individual adsorbed atoms. This information is of
course complicated by the multiple scattering of the incident
particle wavefunction by the adsorbed atoms and the atoms of the
solid substrate. However if there is two-dimensional periodicity
in the adsorbed layer (usually as a consequence of the crystal
symmetry of the substrate), the multiple scattering calculations
of the scattering intensity are significantly simplified.[2,3] Because of the weakness of the van der Waals' image force responsible for the physisorbtion bond, it is difficult to produce an
ordered absorbed layer. An ordered layer is considerably easier
in the case of chemisorbtion where the adsorbed atom is chemically
bonded to the substrate. The added complexity in this case is the
radical distortion of the chemisorbed atom's electronic wavefunction.[4] The goal of studying particle diffraction from crystal
surfaces, with or without ordered adsorbed layers, is to determine
the atomic structure of the surface and to gain understanding of
the bonding of atoms to surfaces. The precision with which these
goals can be achieved is of course dependent on our understanding
of positron and Ps scattering by atoms at surfaces. In this talk
I will touch on some of the advantages and experimental considera-

tions pertinent to low energy positron diffraction LEPD and low
energy Ps diffraction LEPSD as alternatives to the well established surface structure diagnostic techniques: low energy electron
diffraction LEED[5] and elastic helium atom diffraction[6] EHAD.

The first LEPD measurements were made in 1979 at Brandeis with
a slow positron beam having a flux of 4×10^4 s^{-1} which could be
focussed to a 6 mm diameter with an angular spread of 8° (RMS) at
an energy of 100 eV, for example.[7] This beam was produced with an
annealed tungsten parallel vane converter,[8] a 400 mCi Co-58
primary positron source, and 20% transmission aperturing. If one
were to take advantage of the most current high performance
converter (single crystal W(110) in the single backscattering
geometry having $\varepsilon = 3 \times 10^{-3}$, diameter = 10 mm, and transverse
energy component $E_T = 0.4$ eV RMS)[9] then it is possible to obtain a
8×10^6 s^{-1} slow positron flux (with 500 mCi of Co-58) which is
focussable at 100 eV into a 80 mm-degree beam diameter-angular
spread product. Aperturing this beam to a flux of 4×10^4 s^{-1}
(a lower limit on flux needed for "practical" LEPD), would reduce
the diameter-angle product to 6 mm-degree. This is still larger
than the 1 mm-degree figure used in high resolution LEED studies.
To improve this situation one can either produce higher fluxes
with future high intensity (>10^8 s^{-1}) slow positron facities and
aperture the beam further or "brightness enhance" the existing
bench-top intense (10^6 -10^7 s^{-1}) beams by means of accellerating
the slow positrons and reconverting them into slow positrons
reemitted from a greatly reduced diameter.[10] Since brightness
enhancement can now be achieved with no new technology, this
approach seems better to pursue for future LEPD measurements than
resorting to high intensity beams.

Until recently, the method thought to be most feasible for
brightness enhancement required clean, defect free, self-supporting,
single-crystal, < 1000 Å thickness foils. Although this is still
considered possïble, it is reassuring to know that there is now an
easier method available to achieve brightness enhancement. Because
of recent improvements in slow positron conversion using the W(110)
single crystal converter, it is now possible to achieve brightness
enhancement with two stages of moderate brightness enhancement in
the reflection mode with ordinary thick W(110) crystals comparable
to that obtained with single stage transmission mode using thin
foils. The original difficulty with the reflection mode concept
was that electrostatic lenses capable of accelerating and focussing
positrons down to a sufficiently small spot (i.e. one-hundreth the
original beam diameter) at the converter surface also produce electric fields which interfere with subsequent extraction of the
reemitted slow positrons from the same surface. If, however, one
chooses to focus the beam down to only a tenth of its original
diameter (with acceleration to 10 keV), then it is a simple matter
to use a two-tube lens with a focal length equal to four tube diam-

eters and achieve neglible electric field at the focus. Independent measurements at Bell Labs and at Brookhaven have established
that W(110) will reemit incident 10 keV positrons as slow positrons
with at least 30% probability and a maximum transverse energy of
0.4 eV.[9] Thus two stages of reducing the beam diameter by a tenth
each time (using W(110) in the reflection mode) guarantees an
increase in brightness-per-volt by 10^3. This amount of brightness
enhancement allows a 1 mm-degree 100 eV beam having a flux of
7×10^5 s^{-1}. Using a microchannel-plate detector array to detect
all of the diffracted positrons will enable a high resolution LEPD
intensity vs. incident energy (I-V) spectrum to be obtained in a
few minutes. Thus we see that high intensity positron facilities
are not yet necessary for the next generation of LEPD studies.

The question of future directions in LEPD and what relevant
role high intensity positron facilities might play also goes hand-
in-hand with the question of what new physics is to be gained with
further LEPD studies. As mentioned earlier, LEPD measurements from
clean and adsorbed systems offer insight into various perturbative
approaches to understanding positron-atom scattering. In the case
of scattering from strongly perturbed atoms at the surface, LEPD
intensities for $E{\gtrsim}30$ eV are not as sensitive to the delicate
balance[11] between core repulsion and induced polarization as is
the case for lower energy (≈ 1 -10 eV) scattering from neutral atoms.
If the goal is to use LEPD as a surface structure tool, then this
insensitivity is a blessing. For very low energy positron diffraction VLEPD, the scattering of positrons in the 1-30 eV range
requires treatment of positron-electron (core as well as valence)
correlation to be included in calculations. This is a considerable
challenge for theorists in the case of chemisorbed atoms. In this
connection, it is important to note that the VLEPD measurements
obtained at Bell Labs from clean metals demonstrate a strong dependence on positron correlation with the conduction electrons of the
metal surface.[12]

The relative insensitivity of LEPD to positron-electron correlation for $E>30$ eV is due both to the velocity of the positron, in
the case of correlation with valence electrons, and the ion core
repulsion of the positron wave function, in the case of correlation
with ion core electrons. Because of the reduced sensitivity of the
positron to correlation interaction with the ion core electrons,
one is able to calculate LEPD intensities with more confidence
than for LEED, particularly in high-Z materials where correlation
effects have limited the precision of LEED as a surface structure
tool.[13] Also, ion-core repulsion for positrons virtually elimates
the need for relativistic corrections (including spin-orbit coupling
with the ion core).[14] This, and other considerations, points out
that LEPD has the potential (positive) to ultimately provide
structural information about ordered surfaces that is more precise
than that obtained from LEED. A less ambitious claim is that LEPD

will be a valuable complement to LEED and that direct comparisons of LEPD and LEED from the same sample[15] will reduce the uncertainty in structural determination to systematic errors, both experimental and theoretical. As the precision of LEPD and LEED is improved, improved calculations of the scattering phase shifts for the adsorbate atoms will become more critical. At this point, the precision of bench-top LEPD will have to be improved; which ultimately will require a high intensity positron beam facility.

In contrast to LEPD which has been demonstrated to be experimentially feasible without high intensity positron beams, one can view LEPSD as being only marginally possible in the near future and probably not even demonstrable without a high intensity positron beam facility. However the potential rewards in carrying out LEPSD warrant at least an attempt to see if it is possible. Since Ps can only undergo elastic reflection from the outer surface layer of a solid, LEPSD would be an ideal probe of surface structure. This is somewhat similar to the situation for EHAD, which is a powerful tool in surface structure determination because it is mainly sensitive to only the outer surface layer.[6] However, the savings in complexity in not having to treat multiple scattering from subsurface layers (as in the case of LEED and LEPD) is somewhat mitigated by having to deal with long range forces that dominate in the diffraction. The ≈ 0.1 eV energies necessary for He atoms to have ≈ 1Å de Broglie wavelength results in the He atoms having classical turn-ing radii far enough from the individual ion cores that the main scattering is due to the average potential presented by the surface. Due to the repulsion between the conduction electrons, which extend beyond the ionic boundary of a metal, and the closed shell electrons of He, incident He atoms are typically reflected at 3-4Å from the boundary.[16] Thus an accurate treatment of the long range average potential (including many body effects) is required for intensity analysis. However in order for Ps to have a ≈ 1Å wavelength, its energy must be of the order of ≈ 75 eV. At this energy, Ps would be insensitive to the mean surface potential and only undergo elastic reflection in close encounters with the ion cores. Because of the large break-up probability of Ps, multiple scattering and other subsurface contributions to the elastically scattered Ps are expected to be neglible. Thus in principle Ps could be the most ideal "particle" for surface diffraction --short range, outer surface sensitivity only. The practical problem, however, is one of obtaining a sufficient flux of collimated Ps and also having a sufficiently large elastic scattering cross-section versus break-up or spin-exchange into para-Ps). The former problem might be best approached using an intense positron beam facility. At the same time it would be important to be armed with order-of-magnitude estimates of the Ps elastic reflection coefficient in the 20-100 eV range. The lower energy limit at which many-body correlation effects become important is not as well established for LEPSD as it is in the case

of LEPD and LEED. Estimates of Ps reflection coefficients would
aid greatly in ascertaining the feasibility of various experimental
designs to observe LEPSD. The simplest system to investigate might
be one or more physisorbed layers of He on a low temperature substrate. This becomes an especially interesting scattering problem
for very slow Ps ($\approx$ KT - 1 eV) scattering if a Ps-He scattering
length can be determined from the reflection coefficient. At higher
energies, the possibility of disappearance of 24.5 eV Ps due to
formation of $(PsHe)^+$, as suggested by Drachman,[17] would also be
exciting. Measuring the energy of the ejected electron in the reaction would greatly aid in observing this effect.

A rough estimate of the fast Ps beams for LEPSD that one may
anticipate being able to produce in the future can be obtained from
consideration of the "beam-foil" approach, in particular passing
$\approx$ 100 eV positrons through the same 30Å carbon foils used by Mills
for the first observation of the Ps^- ion.[18] Based on two electron
capture rates (i.e. Ps^- formation), the fraction of excited state Ps
formation for backscattered positrons,[19] and comparisons (velocity
scaled) with muonium formation in muonium beam-foil measurements,[20]
efficiencies of 10^{-3}-10^{-2} for forming fast Ps with 100 eV positrons
incident on a 30Å carbon foil is estimated. Fortuntately, some
forward collimation of the emitted Ps should be observed. A limit
on the width of the emitted Ps angular distribution can be obtained
from P_F/P_O, where P_F is the Fermi momentum of the electrons captured by the positrons passing through the foil and P_O is the magnitude of the escaping positronium momentum. For 100 eV Ps, this corresponds to an angular spread of 25° FWHM. Diffusion considerations favor a more collimated emission, i.e. $\cos\theta$.[20] Also we
should take into account the possibility that the electron capture
probability is maximum when the electron is travelling initially
parallel with the positron. A calculation of the Ps angular distribution resulting from formation above the Ps formation threshold
with neutral atoms would be valuable in assesing this conjecture.
The angular spread of hydrogen formed by fast protons of velocities a few times that of outer electron orbital velocities is known
to be $\gtrsim$ 2 times wider than the ratio of the captured electron to
incident proton momenta.[21] This is due to the double scattering
processes involved in electron capture by fast protons and it is
not clear if this process would dominate in the case of the very
light positron. In summary, aperturing a 25° FWHM Ps beam to 5°
FWHM at 100 eV for measuring LEPSD intensities would result in a Ps
flux of 4×10^3 s^{-1}, on target, using an initial slow positron beam
of 10^8 s^{-1}, for example. If the 100 eV Ps reflection coefficient is
$\gtrsim$ 10^{-3} (comparable to that for electrons and positrons), then a
demonstration of LEPSD is possible with first generation high intensity positron beams now being pursued. Beyond this, higher flux
positron beams and more efficient methods of producing collimated
fast Ps are needed to use LEPSD for practical surface determination.

ACKNOWLEDGEMENTS

The author gratefully acknowledges the valuable discussions with Mark Cardillo regarding He atom diffraction and the possible advantages of LEPSD. Also, helpful discussions with Pierro Sferlazzo on positron transmission through thin foils are acknowledged. Support by National Science Foundation grants DMR 8109509 and PHY 8208764, during the preparation of this talk is also acknowledged.

REFERENCES

1. U. Landmann and G. G. Kleiman, Mircroscopic approaches to physisorbtion: theoretical and experimental results, in: "Surface and Defect Properties of Solids, Vol. 6," Chemical Society, London (1976).
2. C. B. Duke, N. O. Lipari, and G. E. Laramore, Surface Structure determination by low-energy electron diffraction, Nuovo Cimento 23: 241 (1974).
3. S. Y. Tong and M. A. Van Hove, Unified comutation scheme of low-energy electron diffraction - the combined-space method, Phys. Rev. B 16: 1459 (1977).
4. N. D. Lang and A. R. Williams, Theory of atomic chemisorbtion on simple metals, Phys. Rev. B 18: 616 (1978).
5. J. B. Pendry, "Low Energy Positron Diffraction", Academic Press, London (1974).
6. H. Hoinkes, The physical interaction potential of gas atoms with single-crystal surfaces, determined from gas-surface diffraction experiments, Rev. Mod. Phys. 52: 933 (1980).
7. I. J. Rosenberg, Alex H. Weiss, and K. F. Canter, Low energy positron diffraction from a Cu(111) surface, Phys. Rev. Lett. 44: 1139 (1980).
8. J. M. Dale, L. D. Hulett, and S. Pendyala, Low energy positrons from metal surfaces, Surf. and Interface Analysis 2: 199 (1980).
9. R. J. Wilson and A. P. Mills, Jr., Clean Single-crystal Tungsten as a positron moderator, Bull. Am. Phys. Soc. 26: 461 (1981).
10. A. P. Mills, Jr., Enhanced brightness of slow positron beams, Appl. Phys. 23: 189 (1980).
11. H. S. W. Massey, E. H. S. Burhop, and H. B. Gilbody, "Electronic and Ionic Impact Phenomena, Vol. 5," Oxford University Press, New York (1975).
12. A. P. Mills, Jr. and P. M. Platzman, Observations of positron Bragg reflection from Al and Cu surfaces, Sol. St. Comm. 35: 321 (1980).
13. M. N. Read and G. J. Russel, On the contraction of the W(001) - (1 × 1) surface using LEED intensity analysis, Surf. Sci. 88: 95 (1979).
14. R. Feder, A theoretical study of low-energy positron diffrac-

tion, Sol. St. Comm. 34: 541 (1980).

15. Alex H. Weiss, I. J. Rosenberg, K. F. Canter, C. B. Duke and A. Paton, Low-energy positron and electron diffraction from Cu(100) and Cu(111), Phys. Rev. B 27: 867 (1983). and Cu(111), Phys. Rev. B 27: 867 (1983).

16. N. Esbjerg and J. K. Nørskov, Dependence of the He-scattering potential on the surface electron-density profile, Phys. Rev. Lett. 45: 807 (1980).

17. R. J. Drachman, Y. K. Ho, and S. K. Houston, Positron attachment to helium in the ^3S state, J. Phys. B 9: L199 (1976).

18. A. P. Mills, Jr., Observation of positronium negative ion, Phys. Rev. Lett. 46: 717 (1981).

19. D. C. Schoepf, S. Berko, K. F. Canter, and Alex H. Weiss, Evidence of excited state positronium formation from metal surfaces in ultra-high vacuum, in: "Positron Annihilation, 6th Conference," P. G. Coleman, S. Sharma, and L. Diana, eds., North-Holland, Amsterdam (1983).

20. P. R. Bolton, A. Badertscher, P.O. Egan, C. J. Gardner, M. Gladisch, V. W. Hughes, J. Vetter, G. zu Pulitz, M. Eckhause, and J. Kane, Observation of muonium in vacuum, Phys. Rev. Lett. 47: 1441 (1981).

21. E. Horsdal-Pedersen, C. L. Cocke, and M. Stockli, Experimental Obeservation of the Thomas Peak in High-Velocity Electron Capture by Protons from He, Phys. Rev. Lett. 50: 1910 (1983).

INDEX

Accelerators, 155–163, 189, 190
  linear accelerators, 155–163, 189
  racetrack microtron, 190
  time bunching, 127, 129–135
  van de Graaff, 190
Adiabatic approximation, 28
Annihilation, see Positron
        annihilation and Positro-
        nium annihilation
  γ-rays from galactic center, 68,
        203

Beams, see Positron beams and
        Positronium beams
Born approximation, 39, 40–47, 49,
        205
  second Born approximation,
        48
Bound states of positrons with
        atoms, 113, 114
Brightness, 122, 123
Brightness enhancement, 121, 125,
        165, 174, 182, 186, 192,
        213, 220

Capacitance manometer, 4, 55
Charge conjugation invariance, 74
Chemical potential, 148
Classical methods in positron–atom
        scattering, 49–50
Close coupling approximation, 42–46
Coincidence parameters, 47
Configuration interaction method,
        111–112
Converters (moderators)
        ^{11}B, 211

Cu, single crystal, 135, 173,
        181, 183
MgO, 66, 181
Pt, 212
W, 55, 181, 183, 195–200, 211,
        212, 220
W, single crystal, 125, 126,
        128, 135, 195–197, 220,
        221
Conversion efficiency, 127, 157,
        161, 166, 167, 181, 211
Correlated configurations method,
        112, 113
Correlations, electron–positron,
        111, 221
C-parity forbidden decays, 74
Crossed beams, 17, 18
Cross sections, see Positron
        scattering and Electron
        scattering

Density fluctuations, 73
Detectors
  channel electron multiplier,
        2, 16, 18, 55, 56
  channel plate, 18, 191
  microchannel plate, 157, 162
  multi-channel plate, 78
  NaI, 55, 157, 215
  plastic scintillator, 2, 3
Differential cross sections, 2,
        15, 215
  e^- – atom scattering, 17, 18,
        45, 47
  e^+ – atom scattering, 15–17,
        41, 45

Diffusion in solids, 142-145, 176
Dispersion relation, 205, 206
Distorted wave method, 46, 49
Doppler free excitation, 77
Drift velocity, 11

Effective range theory, 11
Eikonal-Born series, 40-46
Elastic scattering, 18, 19, 27, 30, 31
  e^+ - Ar, 41, 57
  e^+ - H, 27, 28, 41, 42, 204
  e^+ - He, 41, 42, 43
  e^+ - Ne, 41
Electron-positron correlations, 111, 221
Electron scattering, 1, 7, 19
  differences between e^- and e^+ scattering, 44, 79
Exchange scattering amplitude, 44
Excitation cross sections, see Inelastic cross sections

Galactic center, 68, 203
Glauber approximation, 40

Hartree-Fock method, 109-111
Hot positronium, 90, 92
Hulthen-Kohn variational method, 28
Hylleraas wave functions, 113

Ice, positronium emission from, 90, 91
Implantation, 142-145
Inelastic cross sections, 15, 18, 22
  excitation, 21-24, 31, 45, 46, 54, 204, 205
  ionization, 21, 24, 45-48, 54, 204, 214
Ionization cross sections, see Inelastic cross sections

Lifetime spectra, 54, 75, 90
  for mixtures of gases, 100-104, 106
Linear accelerators, 155-163, 187-189
Liouville's theorem, 122, 212

Low energy positron diffraction, 124, 140-142, 176, 219-222
Low energy positronium diffraction, 219, 22, 223
Lyman - α radiation from positronium, 67

Magnetic bottle, 127, 129, 130-134
Magnetic quenching, 75, 76
Moderators, see Converters
Moment T-matrix method, 42
Momentum transfer cross-sections, 11, 54
Muon, 29
Muonium, 223

Negative work function, 140, 147 149

Optically active molecules, 72
Optical potential, 28
Optical model, 41-43
Optical theorem, 43
Ore gap, 66
Ore model of positronium formation, 66, 85-90
Ortho-positronium, 20, 57 (see also Positronium)
  quenching parameter ($_1 Z_{eff}$), 104

Para-positronium, 20 (see also Positronium)
Polarization potential, 28, 30-34
Polarized density method, 30
Polarized orbital method, 31, 111
Polarized positron beams, 68, 71, 72, 141, 149, 170, 177
Positron annihilation
  annihilation rates and lifetimes, 109-118
  in helium, 207
  in solids, 145
Positron beams, 2-7, 17, 18, 122, 136, 156, 184
  intense beams, 121-124, 165-167, 181-192, 211

using linear accelerators, 155–
163, 167, 169, 187–189
using reactors, 166–172, 186
polarized beams, see Polarized
positron beams
Positron scattering, 1–11, 15–24,
27–35
by atoms
H, 10, 43, 70, 216
He, 44, 61
K, 10
rare gases, 8, 15, 16, 19, 57,
61
by molecules, 7–9, 32–35, 45
differences between e^+ and e^-
cross sections, 44, 79
differential cross sections,
15–17, 41, 45
Positronium, 65–81
annihilation, 204
in gas mixtures, 101
rate, 101
energy distribution of γ-rays,
77
excited states, 61, 65, 67, 80,
205, 223
lifetime and decay rate, 73–77
ortho-positronium, 20, 57
para-positronium, 20, 74, 75
spectroscopy, 77, 78
spin state selection, 68
thermal, 67, 140
Positronium – atom complexes, 66
Positronium beams, 67
Positronium formation
in gases, 10, 18–20, 48, 53–58,
85, 89, 90, 100, 214, 223
H atom 28, 29, 49, 204, 205,
216
H^-, 206
molecules, 20, 50, 55, 60, 61
rare gases, 20, 55, 59–61, 86
inhibition of, 90
in liquids, 85, 86, 89, 90
in powders, 66
at solid surfaces, 67, 90, 140,
149
time, 96
Positronium fraction, 100, 101, 104

dependence on density, 103
Positronium hydride, 111, 112,
206, 207
Positronium molecule, 124
Positronium negative ion, 65–68
76, 141
Positronium – surface inter-
actions, 173
Pseudostates, 45

QED, tests of, 80
Quality factor, 27, 28

Radioactive corrections, 74, 75
80
Radiator-converter, 156–158
Ramsauer effect, 8, 30
Re-emission yield of positrons
and positronium, 146
Resonances, 23, 24, 29

Second order potential method,
41–43
Slow positrons, 2, 15, 147, 148,
187 (see also Positron
beams)
Source-converter configuration,
185
Sources, 168, 182–190, 197–199,
212, 213
^{11}C, 4, 167, 187, 199, 200, 212
^{58}Co, 127, 132, 156, 166, 181–
183, 197, 212
^{64}Cu, 121, 169, 170, 186, 212
^{68}Ge, 76
^{22}Na, 3, 5, 6, 76, 77, 156, 166,
183, 197, 199, 212
Spectrometers, 2–6
Spur model of positronium formation, 66, 85, 88–90
Sum rule, 44
Surface defects, 69, 71
Surface magnetism, 71, 170
Surface states, 140, 141, 149–151,
173

Thermal desorption of positronium,
67
Thermalization distance, 93, 94
Thermalization time, 92

Thermalization in solids, 143–145

Thermalized positrons, 140

Time bunching, 161
  debunching, 169

Time bunching accelerators, see
        Accelerators

Time of flight methods, 1–3, 15,
        21, 54–56

Total cross sections, see Positron
        scattering and Electron
        scattering

Z_{eff}, 32, 34, 101–104 (see also
        Positron annihilation)
  effect of density, 105–107
  effect of temperature, 104–107

$_1Z_{eff}$, 104